ENHANCING THE COMMUNITY COLLEGE PATHWAY TO ENGINEERING CAREERS

Mary C. Mattis and John Sislin, Editors

Committee on Enhancing the Community College Pathway to Engineering Careers

Committee on Engineering Education
National Academy of Engineering

Board on Higher Education and Workforce
Policy and Global Affairs

NATIONAL ACADEMY OF ENGINEERING *AND*
NATIONAL RESEARCH COUNCIL
OF THE NATIONAL ACADEMIES

THE NATIONAL ACADEMIES PRESS
Washington, D.C.
www.nap.edu

THE NATIONAL ACADEMIES PRESS 500 Fifth Street, N.W. Washington, DC 20001

This project was supported by NSF Award No. DUE 0343405 between the National Science Foundation and the National Academy of Sciences. Any opinions, findings, conclusions, or recommendations expressed in this publication are those of the author(s) and do not necessarily reflect the views of the organizations or agencies that provided support for the project.

International Standard Book Number 0-309-09534-4 (Book)
International Standard Book Number 0-309-54757-1 (PDF)
Library of Congress Control Number 2005934050

Additional copies of this report are available from the National Academies Press, 500 Fifth Street, N.W., Lockbox 285, Washington, DC 20055; (800) 624-6242 or (202) 334-3313 (in the Washington metropolitan area); Internet, http://www.nap.edu.

Printed in the United States of America

THE NATIONAL ACADEMIES

Advisers to the Nation on Science, Engineering, and Medicine

The **National Academy of Sciences** is a private, nonprofit, self-perpetuating society of distinguished scholars engaged in scientific and engineering research, dedicated to the furtherance of science and technology and to their use for the general welfare. Upon the authority of the charter granted to it by the Congress in 1863, the Academy has a mandate that requires it to advise the federal government on scientific and technical matters. Dr. Ralph Cicerone is president of the National Academy of Sciences.

The **National Academy of Engineering** was established in 1964, under the charter of the National Academy of Sciences, as a parallel organization of outstanding engineers. It is autonomous in its administration and in the selection of its members, sharing with the National Academy of Sciences the responsibility for advising the federal government. The National Academy of Engineering also sponsors engineering programs aimed at meeting national needs, encourages education and research, and recognizes the superior achievements of engineers. Dr. Wm. A. Wulf is president of the National Academy of Engineering.

The **Institute of Medicine** was established in 1970 by the National Academy of Sciences to secure the services of eminent members of appropriate professions in the examination of policy matters pertaining to the health of the public. The Institute acts under the responsibility given to the National Academy of Sciences by its congressional charter to be an adviser to the federal government and, upon its own initiative, to identify issues of medical care, research, and education. Dr. Harvey V. Fineberg is president of the Institute of Medicine.

The **National Research Council** was organized by the National Academy of Sciences in 1916 to associate the broad community of science and technology with the Academy's purposes of furthering knowledge and advising the federal government. Functioning in accordance with general policies determined by the Academy, the Council has become the principal operating agency of both the National Academy of Sciences and the National Academy of Engineering in providing services to the government, the public, and the scientific and engineering communities. The Council is administered jointly by both Academies and the Institute of Medicine. Dr. Ralph Cicerone and Dr. Wm. A. Wulf are chair and vice chair, respectively, of the National Research Council.

www.national-academies.org

COMMITTEE ON ENHANCING THE COMMUNITY COLLEGE PATHWAY TO ENGINEERING CAREERS

James M. Rosser (Chair), President, California State University, Los Angeles
Ashok Agrawal, Dean, Math, Science Engineering, and Technology, St. Louis Community College
Warren Baker, President, California Polytechnic State University, San Luis Obispo
Richard Culver, Bartle Professor in Mechanical Engineering, Binghamton University, State University of New York
Dan Dimitriu, Engineering Program Coordinator, San Antonio College
Jack R. Lohmann, Associate Provost, Georgia Institute of Technology
Margaret Weeks, Adjunct Program Manager, ABET Inc.
Aaron Wenger, Professor Emeritus, Itasca Community College
Vera Zdravkovich, Vice President for Instruction, Prince Georges Community College

National Academy of Engineering Staff

Patricia Mead, Study Director *(until May 2004)*
Mary Mattis, Study Director *(from May 2004)*
Nathan Kahl, Senior Project Assistant
Carol R. Arenberg, NAE Senior Editor

National Research Council Staff

Peter Henderson, Director, Board on Higher Education and Workforce
John Sislin, Program Officer

Preface

The important role of community colleges in educating engineers is not well known to the public, or even to the engineering community. In fact, 20 percent of engineering degree holders began their academic careers with at least 10 credits from community colleges, and 40 percent of the recipients of engineering bachelor and masters degrees in 1999 and 2000 attended community colleges. In addition, community colleges offer unique opportunities for increasing diversity in the engineering workforce.

The role of community colleges in engineering education could be expanded if a number of steps were taken to improve transfer partnerships with four-year engineering programs. This report describes how some community colleges and four-year educational institutions have facilitated the transfer process for students transferring from two-year engineering-science programs to four-year engineering programs. The report also provides recommendations for improving transfers overall and otherwise enhancing the role of community colleges in educating engineers.

At the request of the National Science Foundation, the president of the National Academy of Engineering and the chair of the National Research Council appointed an ad hoc committee to design and oversee the study. The committee met four times in person and by teleconference to develop the study design and to hear expert testimony. The primary fact-finding activity was a workshop exploring the exemplary approaches of 24 transfer partnerships between two- and four-year colleges. The focus of the workshop was on transfer and articulation (policies and programs designed to foster transfer); recruitment and retention; curriculum, quality, and standards; diversity; and data collection. The committee's deliberations, reflected in this report, were based on the results of the workshop, expert testimony, and a survey of the relevant literature. The final document also reflects the personal and professional experience and judgment of committee members. This study, funded by the National Science Foundation, was conceived as the first phase of a two-part project. Phase II, which is in the planning stages, will focus on one or more questions for future research identified in this report.

Many students, especially students from groups that are underrepresented in the engineering workforce, are increasingly turning to community colleges as a starting point for their postsecondary education. Yet many of them are not aware that they can earn a B.S. in engineering through the community college pathway. Informing middle-school and high-school students and their parents of the opportunities offered by community colleges will require a major campaign by community colleges, four-year educational institutions, engineering societies, and the state and federal agencies responsible for educational outcomes. Students and parents must be given assurances that community colleges and four-year programs will work together to provide seamless transfers from two- to four-year programs and offer a supportive educational environment that promotes the retention of students working toward [associate of science (A.S.) and bachelor of science (B.S.)] degrees in engineering. This report describes concrete examples of activities undertaken by community colleges and four-year educational institutions in support of these goals.

James M. Rosser, *Chair*
Committee on Enhancing the Community
College Pathway to Engineering Careers

Acknowledgments

This report has been reviewed in draft form by individuals chosen for their diverse perspectives and technical expertise, in accordance with procedures approved by the NRC's Report Review Committee. The purpose of this independent review is to provide candid and critical comments that will assist the institution in making its published report as sound as possible and to ensure that the report meets institutional standards for objectivity, evidence, and responsiveness to the study charge. The review comments and draft manuscript remain confidential to protect the integrity of the deliberative process.

We wish to thank the following individuals for their review of this report: George Boggs, American Association of Community Colleges; Roger Bowen, American Association of University Professors; George Campbell, Jr., Cooper Union for the Advancement of Science; Alfredo de los Santos, Arizona State University; Michael Gibbons, American Society for Engineering Education; Susan Hackwood, University of California at Riverside; Chen-Ching Liu, University of Washington; John Wadach, Monroe Community College; Jane Wellman, Institute for Higher Education Policy; and James West, Johns Hopkins University.

Although the reviewers listed above have provided many constructive comments and suggestions, they were not asked to endorse the conclusions or recommendations, nor did they see the final draft of the report before its release. The review of this report was overseen by Gerald Dinneen, Consultant. Appointed by the National Research Council, he was responsible for making certain that an independent examination of this report was carried out in accordance with institutional procedures and that all review comments were carefully considered. Responsibility for the final content of this report rests entirely with the authoring committee and the institution.

The production of this report was the result of work by the study committee over a sustained period of time. The committee was ably assisted by Patricia Mead, study director during the first part of the study; Mary Mattis, study director during the second half of the project; Peter Henderson, director of the Board on Higher Education and Workforce; John Sislin, program officer; and Nathan Kahl, senior project assistant.

Contents

Summary

This study, funded by the National Science Foundation, was conceived as the first phase of a two-part project. For the first phase of the study, the findings of which are presented in this report, the committee was asked to (1) describe the changing role of community colleges in engineering education; (2) identify exemplary practices and partnerships between community colleges and four-year educational institutions; and (3) recommend critical areas for further study. The expected second phase of the study, to take place at a later time, would focus on selected research questions and provide guidance for policy makers. The study committee was overseen by the Committee on Engineering Education and the Committee on Diversity in the Engineering Workforce, both of the National Academy of Engineering, and the Board on Higher Education and Workforce of the National Research Council.

The committee held two meetings that included presentations by outside experts, reviews of the literature, and guided discussions and deliberations; organized and conducted a two-day workshop that brought together selected administrators of community colleges and four-year institutions and faculty members from notable programs; and collected and analyzed data on recent engineering graduates with community college experience.

Community colleges are already essential to the education of engineers in the United States. Indeed, 20 percent of engineering degree holders began their academic careers starting in and earning at least 10 credits at community colleges (Adelman, 1998), and 40 percent of the recipients of engineering bachelor's and master's degrees in 1999 and 2000 attended community colleges at some time (Tsapogas, 2004). Nevertheless, the committee believes that community colleges have not achieved their full potential for several reasons: (1) a lack of understanding among parents, teachers, counselors, and students of the effectiveness of community colleges in producing engineering graduates; (2) less than effective articulation agreements (policies and programs designed to foster transfer) between community colleges and four-year institutions; and (3) a lack of cooperation and coordination among high schools, community colleges, four-year institutions, and state higher-education agencies.

The workshop focused on five themes, in addition to identifying areas for further research:

1. Challenges and opportunities for improving articulation and transfer between community colleges and four-year educational institutions.
2. The recruitment and retention of students at various junctures of the community college pathway to engineering careers.
3. The curricular content, quality, and standards of two-year A.S. programs and of four-year engineering programs.
4. Opportunities for community colleges to increase diversity in the engineering workforce.
5. Sources of data on community college and transfer students and the need for more systematic data collection.

The report provides descriptions of exemplary programs and practices of community colleges and four-year educational institutions; outreach activities designed to recruit and retain K–12 students through the completion of the baccalaureate degree; and statewide initiatives focused on articulation and transfer. The committee recognizes that there is no "one size fits all" approach to articulation and transfer programs; therefore, the workshop was designed to identify a variety of ways community colleges and four-year educational institutions could improve pathways to careers in engineering and improve educational outcomes in preparing students to pursue engineering education. Based on the personal and professional experience of committee members and the workshop, the practices detailed in this report were identified as initiatives that have enhanced community college pathways to engineering.

The experiences described by workshop representatives of two- and four-year schools indicate that articulation agreements are necessary, but not sufficient, for seamless transfers of community college students. The committee defined a "good" transfer partnership as a "second-level articulation," that is, cooperative efforts by the two-year and four-year college to recruit students into engineering. Articulation, therefore, should be based on student outcomes and competencies according to ABET guidelines (rather than on course credits, curricula, or the sequences of courses) determined by the university-community college partnership.

Communication between two- and four-year transfer partners is critical to second-level articulation. According to workshop participants, successful transfer partners communicate frequently, visit each other's campuses, meet frequently to discuss curricular changes, and even share faculties. Currently, communication between two-year and four-year faculty members varies from campus to campus and department to department, often depending on personal relationships among faculty members or administrators. Few formal approaches to communication were described.

As the trends in engineering education move toward greater diversity and specialization in the lower division course offerings of four-year engineering programs, engineering science curricula at community colleges are less likely to cover the same material or achieve the same results. Thus, the need for communication and resource sharing between transfer partners and for the timely updating of articulation agreements is becoming more urgent.

Better articulation also requires cooperation between two- and four-year transfer partners in the recruitment of engineering students. Four-year institutions can promote and support the community college pathway as a viable, even attractive, route to a baccalaureate degree in engineering. Individualized counseling and coaching can be provided early and often to students in both types of institutions. Workshop participants indicated that transfer students are often the best recruiters, mentors, and tutors for students at their two-year alma maters because they know how the system works. Ideally, faculty members at both institutions know each other well and, in the best partnerships, collaborate on projects, curriculum development, recruitment, and other activities that support community college transfer students (Rifkin, 1998).

Exemplary articulation and transfer initiatives are characterized by clear, accessible information for parents and students who are candidates for community college engineering science programs, and transparent, accessible documentation regarding the transfer mission between partners. To increase the number of students who embark on the community college pathway to engineering, four-year schools will have to use their brand images to promote community college programs, perhaps by developing joint admission and recruitment programs with two-year schools. For example, the name of a four-year school could be listed next to the engineering science major on the community college application and promotional materials and joint high school outreach programs could be developed.

Workshop participants generally agreed that strong partnerships between community colleges and four-year engineering programs improve student recruitment and retention at both institutions. Community colleges that reach out to potential students through a variety of messages and media and demonstrate that they have a proven record of success in preparing students to transfer to an engineering degree program are more likely to succeed in their recruitment and retention activities, especially if they have an established articulation agreement with a four-year institution. Four-year institutional partners also benefit by being able to draw on an expanded, and in some cases more diverse, recruitment pool that includes talented community college students. Moreover, the retention rates at four-year institutions will improve when two- and four-year programs work together to prepare students to take upper-division engineering courses.

Workshop participants reported that the lack of financial assistance from institutional, state, and federal sources is a significant barrier to the recruitment and retention of engineering science students in community colleges. Students who transfer to four-year engineering programs also need financial assistance to ensure that they can afford to stay in school until graduation

Many workshop participants pointed out a need for better evaluations of articulation agreements and transfer processes. Evaluations should include: definitions of positive outcomes of diversity; assessments of learning outcomes; and assessments of shared-learning, outcomes-based objectives. The main concern expressed at the workshop was that articulation agreements may focus only on easily measured outcomes, such as the number of credits or courses, rather than on important learning outcomes and competencies. In fact, there is a fundamental disagreement between two-year and four-year programs about what evaluations should measure. Another serious problem is that many educational institutions do not collect or analyze data on students that would support assessments; even when data are collected, this is not done uniformly.

Areas for future research include documenting performance outcomes in terms of recruitment, transfer, retention, and persistence to degrees in undergraduate engineering education. Data collected by educational institutions and research organizations on community college student outcomes in engineering education would provide a basis for improving transfer partnerships and articulation agreements

Community colleges have long been recognized as providing opportunities to increase diversity in the U.S. engineering workforce, especially racial and ethnic diversity. Although the makeup of community colleges' student bodies varies by geographic location, a larger percentage of students from some minority groups, notably Hispanics and American Indians, attend community colleges than white, African-American, and Asian students. In effect, community colleges have become an educational pipeline for underrepresented minorities entering the higher education system.

Organizations in the engineering educational and professional communities could work together to increase the awareness of the importance of diversity in the engineering workforce and to educate state and federal legislators. State and federal funding for community college students and incentives for four-year engineering institutions to reach out to community colleges and their students could eventually lead to increases in the number of underrepresented minorities in engineering.

Another area for future activity is the collection of comparative data to identify factors associated with the retention and persistence to the B.S. degree of women, minority, and nonminority male community college and transfer students.

The lack of information (especially longitudinal and comparative information that can be disaggregated by gender, race/ethnicity, and other background variables) on the successes and failures of students who begin their engineering educations in two-year and four-year programs presents serious problems for an analysis of the transfer function of community colleges. Most often, community colleges lose sight of students once they transfer to four-year institutions, precisely when they should begin tracking their educational and career trajectories. Compiling and publicizing data on transfer students' success in obtaining B.S. or advanced engineering degrees would demonstrate the effectiveness of engineering studies in community colleges and improve their recruitment rates.

A comprehensive, systematic strategy for data collection on educational and career outcomes for community college and transfer students would require leadership in the engineering profession and from funding agencies to define the most relevant data items, to encourage collaboration between two- and four-year educational institutions, to ensure the privacy of students, and to develop vehicles for dissemination.

Finally, the committee notes that the engineering education community, and the profession as a whole, would benefit from further discussion of the feasibility and desirability of accreditation for engineering science programs at community colleges.

AREAS FOR FUTURE RESEARCH

The committee identified a number of questions for further research:

- What is the attrition rate of students who begin their engineering studies at community colleges?

- Do engineering students who begin at community colleges perform as well, better than, or not as well as other students? (A considerable amount of *anecdotal* information on this question was provided by workshop participants.)
- What factors in the culture, student services, and learning environments of community colleges and four-year engineering programs correlate with the retention and persistence of students to the B.S. degree?
- What evidence is there that community college engineering science students learn effectively via online courses?
- How can the teaching of mathematics be more focused on engineering applications?
- What competencies should students have after two years in engineering science programs?
- What data would persuade faculty and administrators of two- and four-year educational institutions and state and federal policy makers to enhance the community college pathway to engineering degrees and careers?

Although this study examines partnerships between community colleges and four-year engineering programs, the primary focus is on the needs of community colleges and their students related to articulation agreements and transfer processes. Research on the perspectives of four-year educational institutions would also be helpful, as would an in-depth examination using both quantitative and qualitative data-collection methods of the experiences of a cohort of students entering and progressing through the community college pathway to engineering careers.

1

Overview

According to the U.S. Congress, building a larger and more diverse workforce educated in science, technology, engineering, and mathematics is a critical national imperative for the twenty-first century (DOD, 2001). Increasing the number of engineers will first require increasing the number of engineering students, and one way to do that is to tap into the pool of students pursuing engineering science studies at community colleges, who could then transfer to four-year institutions, where they could pursue baccalaureate or advanced degrees. Community colleges are an especially attractive source of prospective engineering students for several reasons: (1) millions of students attend them, and enrollments are projected to increase (DOED, 2004); (2) many women and students from underrepresented minorities attend community colleges; and most important, (3) many community college students in engineering do not transfer to four-year engineering programs. Nevertheless, community college transfer students who have completed an associate of science (A.S.) degree in engineering science are just as likely to receive a bachelor's degree as students who attend four-year campuses only. Therefore, it makes sense for educators, legislators, industry, and other stakeholders to pay more attention to the potential talent pool available at community colleges (Adelman, 1998).

Community colleges are already essential to the education of engineers in the United States. A U.S. Department of Education study reported that 20.1 percent of engineering degree holders began their academic careers with at least 10 credits from community colleges (Adelman, 1998), and 40 percent of engineering bachelor's and master's degree recipients in

1999 and 2000 attended community colleges (Tsapogas, 2004). In some regions of the country, the numbers are even larger. The California Council on Science and Technology reported that 48 percent of graduates with science or engineering degrees from the California system began at community colleges and then transferred (CCST, 2002).

However, among the various functions of community colleges, the transfer function has yet to reach its full potential, for many reasons (Ulseth and Wenger, 2002):

- a lack of understanding on the part of parents, teachers, counselors, and students of the importance of community colleges in engineering education
- disagreement over curricular content and quality
- ineffective articulation agreements
- a lack of cooperation between high schools, community colleges, four-year institutions, and state higher education agencies
- inadequate attention to the transfer of women and underrepresented minority students
- insufficient statistics on student transfers, experiences, and outcomes

CHARGE TO THE COMMITTEE

The study committee was charged with describing the evolving role of community colleges in engineering education, identifying exemplary programs at community colleges and model partnerships between two-year and four-year institutions, and recommending critical research questions that require further study. The study committee was overseen by the National Academy of Engineering (NAE) Committee on Diversity in the Engineering Workforce and Committee on Engineering Education and the National Research Council (NRC) Board on Higher Education and Workforce.

METHODOLOGY

The NAE president and NRC chair appointed an ad hoc committee to carry out this study during 2004. Most of the committee's time and effort involved carrying out fact-finding activities and analyzing the collected information. The main fact-finding activity was a one and one-half day workshop described below.

The members of the study committee, who were chosen for their expertise and experience, provided the insight and analysis for this report, which is intended to initiate a national dialogue on enhancing the role of community colleges in undergraduate engineering education. Committee members included: community-college and four-year engineering educators; experts on administrative processes and articulation agreements between community colleges and four-year campuses; higher education program evaluators; and researchers who have studied the demographics of community college students and the impact of community colleges on the development of the science, technology, engineering, and mathematics workforce. Biographies of committee members can be found in Appendix A.

During the fact-finding stage of this study, the committee gathered evidence to address its charge through the following activities:

- two committee meetings with presentations by outside experts, reviews of background research, and guided discussions and deliberations
- a one and one-half day workshop, which brought together two- and four-year college administrators and faculty members from notable programs
- the collection and analysis of data on community college involvement by recent engineering graduates

The study committee met on three occasions during 2004—a teleconference planning meeting on March 8, a full committee meeting in Washington, D.C., on April 1 (see agenda in Appendix B), and a full committee meeting on July 8 and 9 (see agenda in Appendix C). The meeting in March included expert testimony from stakeholders and planning for the workshop. Outside experts discussed national and state perspectives on the community college transfer mission; curricular reform and diversity issues; community college/four-year engineering program partnerships; and the need for a national certification of competencies necessary for student success in four-year programs. The third committee meeting was focused on preparing the report.

The workshop was held on July 7 and 8 in Washington, D.C., to showcase programs that have demonstrated their ability to produce successful engineering graduates with community college experience. Keynote speakers and experts, who participated in four panel discussions, described their experiences as faculty members or administrators of community colleges or four-year educational institutions and in state organizations that develop and evaluate educational initiatives. Workshop participants were asked to draw on their own experiences to suggest questions relating to the transfer mission that would require additional research. The agenda for the workshop is in Appendix D.

In the notable programs profiled during the workshop, proactive planning has resulted in successful partnerships between community college and four-year engineering programs. These partnerships were focused on the following factors related to articulation and transfer:

- the success of students who transfer between community colleges and four-year engineering programs
- mutually beneficial interaction between faculties of community colleges and four-year engineering programs
- the curricular content of community college engineering programs
- data-collection requirements for effective program evaluation
- effective use of community college campuses to achieve student and faculty diversity goals
- a quality-assurance standard

Comments by the presenters and the general discussion at the workshop are reflected in the research questions presented in subsequent chapters.

Workshop on Key Issues and Exemplary Practices in Community College Engineering Science Programs and Transfer

Because it was not possible to survey the universe of community colleges to select programs to profile at the workshop, educational institutions involved in transfer partnerships were pre-selected and asked to submit proposals for participation in the workshop. The pre-selected programs were chosen from the following sources: committee members' personal knowledge of programs at two- and four-year colleges and universities; public presentations by institutional representatives at the 2002 National Science Foundation (NSF)-sponsored conference—A New Direction in Engineering Education: Creating a National Collaboration—and other events; institutions that were represented on the American Society for Engineering Education (ASEE) two-year college division listserv; and institutions represented by members of the League Alliance of the League for Innovation in the Community. Responses were received from faculty and/or administrators from community colleges and four-year educational institutions representing 24 transfer partnerships (in some cases more than two institutions were involved in a partnership). Follow-up communications included requests for additional program information and outcome date, where available.

As noted elsewhere in this report, most two-year institutions do not have the resources to compile and analyze data on the effectiveness of their preparation of engineering science students for transfer to four-year engineering programs. Therefore, the initiatives profiled in this report were selected largely on the basis of qualitative data provided in written proposals and workshop presentations, as well as the personal and professional experiences and judgments of committee members. In addition, because the committee was aware that there is no "one size fits all" approach to articulation and transfer, a variety of options was included for both two-year and four-year educational institutions to enhance community college pathways and improve their preparation of students for careers in engineering.

Faculty members and administrators from 17 four-year educational institutions and 18 two-year educational institutions attended the workshop held on July 7 and 8, 2004. Attendees were from 23 states and the District of Columbia and represented all geographic regions of the United States. A list of workshop participants is provided in Appendix D.

These institutions varied in terms of the racial/ethnic and gender composition of their student bodies and the variety of approaches to their transfer mission. For example, the amount of support provided to transfer students,(academic, financial, counseling before and after transfer) and assistance in job placement on completion of the degree, varied considerably among four-year institutions.

Although all of the institutions represented at the workshop had formal articulation agreements with two-year or four-year institutional partners, there were as many weaknesses as strengths reported in both institutional and statewide agreements. In keeping with information in the literature on transfer students, a majority of representatives of four-year institutions reported that transfer students were "as well prepared" or "better prepared" than four-year-only students and that the retention rate for transfer students was "the same" or "higher" than for four-year-only students.

CONCLUSION

The community college transfer function is critical to meeting the national need for a robust, diverse engineering workforce. In fact, community college transfer may be the primary mechanism for increasing the number of students pursuing engineering degrees, particularly underrepresented minority students. Although there are many innovative and effective partnerships between community colleges and four-year institutions, they must not only be enhanced, but also taken to scale to meet a national need.

This study is a survey of the important questions that must be addressed in future research to promote significant improvements in the community college transfer mission. The following chapters focus on five major areas that could be improved:

1. Articulation agreements to facilitate seamless transfers of students to four-year institutions.
2. Recruitment and retention of engineering students by leveraging the special position of community colleges as "colleges within communities."
3. Curricular content, quality, and standards.
4. Diversity.
5. Data collection.

In each chapter the committee describes the topic, identifies key challenges to improvement, describes notable programs developed by two-year and four-year institutions, and provides conclusions and recommendations for future research. Four overarching themes are highlighted:

1. Shifting program assessments from inputs to learning outcomes.
2. The importance of collaboration.
3. Diversity.
4. Raising awareness of the community college transfer mission.

In the past several years, a number of studies have been published and conferences held to address barriers to improving the transfer function of community colleges. The American Association of Community Colleges and the American Association of State Colleges and Universities, for example, surveyed their institutions and identified the following barriers (AACC and AASCU, 2004):[1]

- the reluctance of four-year institutions to accept credits from associate in applied science (AAS) degree programs
- the lack of child care at four-year institutions
- unavailability of courses at times convenient to transfer and nontraditional students
- the lack of financial aid packages structured specifically for adult and independent students
- admission requirements in programs at four-year institutions that exceed general education requirements for those institutions

[1]This report provides detailed discussions of a variety of barriers to access by community college students to baccalaureate degrees.

For community college students, especially for nontraditional students, the institutional capacity, faculty attitudes, lack of advisors and student support, and statewide and system-wide policy barriers can all be roadblocks to successful transfer to four-year engineering programs.

This report affirms the existence of these challenges, highlights innovative and interesting solutions tried by community colleges and four-year institutions to overcome them, and notes that many important research questions have not been answered.

2

Successful Transfers to Four-Year Institutions

The goal of a transfer program in engineering sciences is to prepare students to enter four-year institutions at the upper-division level where they can take specialized coursework for matriculation with bachelor's degrees in engineering. The goal of the two-year institution is for the transfer to be seamless. However, many four-year institutions present obstacles to seamless transfers, including weak, or even non-binding, articulation agreements and a lack of collaboration and communication between the four-year institution and the community college.

KEY CHALLENGES

Dimitriu and O'Connor (2004) identify four vital elements to recruiting and retaining students in community college engineering science programs and preparing them to succeed after transfer to a four-year university:

1. An aggressive high school outreach program to attract qualified students to the engineering profession via community colleges and four-year educational institutions.
2. An intensive enrichment program in mathematics and science at the community college level to boost the proficiency of underprepared high school students.
3. Engineering courses with state-of-the-art technology and education supplemented with a variety of interesting design-and-build programs to motivate students to continue the study of engineering leading to a bachelor's degree.

4. Coordinated curricula between community colleges and four-year institutions through articulation agreements with institutions in the area.

Articulation agreements are often the most prominent feature of the transfer mission. Although more and more states are becoming involved in articulation agreements, there is currently no single model (Rifkin, 1998). Ignash and Towsend (2000) describes some different approaches:

> To ensure ease of transfer, defined by Kintzer and Wattenbarger (1985, p. iii) as "the mechanics of credit, course, and curriculum exchange," most states have emphasized the "documents approach" (Bers, 1994, p. 249). This approach "emphasizes the development and ongoing maintenance of formal or official agreements related to course equivalencies, articulated 2 + 2 programs, legislative or state agency policies related to transfer, and perhaps statistical reports about student transfer, persistence, and academic performance" (Bers, 1994, p. 249). In a deregulated state system, individual institutions may have the responsibility for establishing articulation agreements concerning which courses, programs, and degrees will transfer from institution to institution. In a more regulated system the state may provide some general guidelines and incentives for institutions to develop these agreements; and in a highly regulated system the state may mandate that the associate of arts degree be accepted at all state institutions, as in Florida, for example.

In a 1999 study Ignash found that 34 states (out of 43 responding to an 11-question survey sent to executive directors of state higher education and community college agencies) had developed statewide articulation agreements. Thirty-three of these 34 agreements were two-year to four-year articulations, suggesting that this type of transfer has received the most attention. The majority of statewide agreements involved public-sector institutions only. Another study, *Transfer and Articulation Policies*, conducted by the Education Commission of the States, lists legislation, cooperative agreements, transfer data reporting, incentives and rewards, statewide articulation guides, common core curricula, and common course numbering for each state (ECS, 2001). Wellman (2002) examined state transfer policies in Arkansas, Florida, New Mexico, New York, North Carolina, and Texas to determine how each state "uses state policy to affect transfer performance, looking at several dimensions of state policy: governance, enrollment planning, academic policies affecting transfer, and data collection and accountability."

These studies raise three questions. First, what is the appropriate scope for an articulation agreement? Students whose geographic mobility is constrained by factors other than education may see little value in agreements that cover a large geographical area if they do not address specific local needs. Second, are some articulation agreements better than others, and if so, under what conditions? And third, should the agreement focus on the transfer of particular courses, a block of courses, or an entire program? Matching course content across institutions is difficult for a number of reasons. Courses at two-year institutions may be numbered the same as the courses at a four-year institution, but questions remain as to whether they have the same content, whether the courses will transfer similarly to different four-year institutions (Rifkin, 1998), whether four-year institutions agree on which lower-division courses students should take, and whether differences in graduation requirements at two-year or four-year institutions will affect transfer students. Answering these and other questions will require extensive research.

Course-by-Course Transfers

Participants in the workshop for this study expressed skepticism about course-by-course articulation agreements. For example, during the workshop presentation by representatives from Mississippi State University and Jones County Junior College, participants pointed out that modifications are ongoing in engineering and engineering-related courses and programs at four-year institutions, and information about these changes takes time to trickle down to the two-year institutions. Along those same lines, other speakers raised similar issues:

- increasing differentiation among four-year engineering curricula
- out-of-date articulation agreements
- equivalency guides that do not accurately reflect changes in four-year engineering curricula
- a lack of uniformity among curricula at four-year educational institutions in the same state
- infrequent communication between transfer partners
- a lack of institutionalization of communication between transfer partners
- a lack of formal mechanisms for reviewing curricular changes statewide
- a limited ability of small community colleges to adjust to curricular changes in four-year programs
- cumbersome, arbitrary credit transfer review processes of four-year institutions where department chairs and other engineering faculty members make decisions about equivalency on a course-by-course basis

The following description by a committee member illustrates some problems transfer students may confront. A community college includes Engineering Circuits I and Engineering Circuits II in its Engineering Science Transfer Program. These courses are well established and are accepted by the two leading state universities, although one requires that transfer students pass a test before granting credit. However, the situation is quite different at two other local universities, which are geographically closest to the community college. One teaches a six-credit combined Circuits I and II course; thus, if a student from the community college takes only one Circuits course, it does not count for transfer to that four-year program. The situation at the other university is similar; Circuits I and II are taught in two separate courses, but a student must either transfer both courses or neither course.

Creating Lasting, Effective Partnerships

In some cases, partnerships have functioned because of personal relationships between faculty members at the partner institutions. Although many educational institutions have transfer offices and dedicated advisors, decisions about which courses will transfer are often handled by registrars or admissions offices that may be far removed from the partnership. Faculty members often have the clearest idea of which courses should transfer, and faculty members are the ones who encourage students to consider transferring to four-year institutions. However, if only one or two faculty members are behind the transfer mission, those faculty members may become indispensable to the functioning of the program. If they stop participating or retire, for example, there could be a major interruption in the transfer function. Thus, there is tension between institutionalizing an articulation agreement and creating an institutional bureaucracy (i.e., setting rules and regulations, involving more institutional actors).

A representative from the National Science Foundation's Engineering Directorate, in a workshop presentation, noted that interaction between faculty members from two-year and four-year institutions should be collaborative rather than competitive. Partnerships must be stable, even in the face of personnel or curricular changes

The experiences described by workshop representatives of two- and four-year schools indicate that articulation agreements are necessary, but not sufficient, for seamless transfers of community college students. The committee defined a "good" transfer partnership as a "second-level articulation," that is, cooperation between the two-year and four-year colleges to recruit students into engineering. Articulation, therefore, should be based on student outcomes, rather than on course credits, curricula, or the sequence of courses taken. These outcomes and competences would be determined by the university-community college partnership according to ABET guidelines.

Communications between two- and four-year transfer partners is also critical to second-level articulation. According to workshop participants, successful transfer partners communicate frequently, visit each other's campuses, meet frequently to discuss curricular changes, and even share faculties. Currently, communication between two-year and four-year faculty members varies from campus to campus and department to department, often depending on personal relationships among faculty members or administrators. Few formal approaches to communication were described.

Better articulation also requires cooperation between two- and four-year transfer partners in the recruitment of engineering students. Four-year institutions can promote and support the community college pathway as a viable, even attractive, route to a baccalaureate degree in engineering. Individualized counseling and coaching can be provided early and often to students in both types of institutions. Workshop participants indicated that transfer students are often the best recruiters, mentors, and tutors for students at their two-year alma maters because they know how the system works. Ideally, faculty members at both institutions know each other well and, in the best partnerships, collaborate on projects, curriculum development, recruitment, and other activities that support community college transfer students (Rifkin, 1998).

Several mechanisms for initiating such partnerships were identified during the workshop. In some cases, the partnership was initiated through a grass-roots approach between a few faculty members at two-year and four-year institutions. In a few cases, faculty members at two-year institutions had received degrees from the four-year institutional partner. Some partnerships were initiated by state legislation.

EXEMPLARY APPROACHES

Representatives of two-year and four-year educational institutions had similar opinions about the necessary components of strong articulation agreements. However, four-year institutions placed a great deal of emphasis on the gatekeeper role of their institutions; their concerns were to maintain standards and promote a uniform approach between partner institutions and to require a relatively high GPA for admission and eligibility for financial assistance. Participants at the workshop suggested several ways of enhancing the transfer mission, based on the strengths of existing articulation agreements: clear and accessible print, Web, and other resources summarizing articulation agreements and the transfer process for students and their parents; regular communication between institutional partners to keep articulation agreements current with curricular changes at four-year educational institutions; an assurance of equivalency; and support for students who transfer to four-year engineering programs.

Accessible Information

Many speakers noted that effective articulation agreements have three characteristics—clarity, transparency, and accessibility. These agreements clearly state which courses are necessary before transfer and how and under what conditions courses and course credits will transfer (e.g., GPA requirements). Uniform course numbering systems and easily accessible equivalency guides were cited as effective ways of summarizing the information that community college students need to select courses and arrange schedules. There was general agreement at the workshop that print resources and other forms of communication should assure students and parents that completing the first two years of engineering education at a community college is both affordable and credible. As one participant stated, "They should provide familiarity and ease of mind for parents." If adequate data are available, the success rate of A.S. graduates in completing bachelor's-degree requirements should be made available to students and parents.

Communication between Institutional Partners

Frequent communication, including regular visits and a transfer counselor at each institution, was considered an essential component of effective articulation agreements and transfer relationships between partner institutions. Frequent communication also demonstrates the desire of both institutions to work together. Community colleges, especially, need regular communication with four-year institutions to adapt their courses to changes in curricula. Frequent communication between faculty members and administrators of two-year and four-year programs also leads to greater cooperation and increases the sensitivity of personnel in four-year programs to the impact of changes in lower-division engineering courses on two-year transfer students. Collaborative activities, such as the joint development of grant proposals and workshops, can also increase the level of cooperation between transfer partners.

Equivalency

Most workshop participants expressed the view that successful articulation agreements promote *uniformity* of curricula at two-year and four-year educational institutions. A strong articulation agreement should enable a community college to provide an engineering science course structure that is nearly identical to the lower-division engineering curriculum of its four-year partner(s). The majority of first- and second-year pre-engineering and engineering courses offered at the community college should be the same as those offered at the four-year institution(s).

Equivalency is paramount to community colleges attempting to maximize the number of credits their students can transfer. However, not all four-year institutions have a single course pathway for the first two years of engineering education for all students (e.g., the curriculum for mechanical engineering in the first two years may differ significantly from the electrical engineering curriculum). Some presenters at the workshop argued that two-year institutions should not be expected to match their curricula exactly to those of four-year engineering programs. Articulation agreements, they suggested, should promote *flexibility*. They could, for example, grant free or elective technical credits to allow transfer students to complete bachelor's degrees in four semesters.

Support for Transfer Students

Chief among the reasons that students use the community college pathway to a four-year degree is the need to minimize the cost of their college educations. Most workshop participants considered financial aid, or equivalent support (e.g., cooperative opportunities), for transfer students an essential component of articulation agreements; yet, many four-year institutions do not offer scholarships or other forms of financial assistance to transfer students. In contrast, exemplary articulation agreements offer transfer students per-semester scholarships, allow students to take courses at a community college with financial aid from the four-year institution, and require a single application process and fee for partnering institutions. The New England Compact, for example, allows students to transfer to out-of-state educational institutions at reduced tuition rates if their home states do not offer higher degree programs in their fields of study. Research in this area could provide a more accurate picture of the availability of financial aid to transfer students.

Institutional Policies that Facilitate Transfer

The Kentucky Council on Postsecondary Education General Education Transfer Policy

This policy was designed to improve the transferability of general education coursework among public colleges and universities in the state of Kentucky. Students who complete a 60-hour program described in the "General Education Block Transfer Policy," earn the Associate of Arts (A.A.) or Associate of Science (A.S.) degree and are certified eligible for admission with junior-level standing to four-year universities. Their general education courses are accepted as meeting up to 48 hours of institution-wide lower-division transfer credits. This may eliminate the need to for a time-consuming course-by-course analysis.

University of Kentucky Extended Campus, Paducah

This partnership between the University of Kentucky four-year engineering program and the West Kentucky Community and Technology College engineering science program has a unique feature—both institutions are located on the same campus. This arrangement greatly facilitates the transfer of students from the two-year to the four-year program because it provides opportunities early on for engineering science students to access information about transfer requirements in an environment that offers faculty and peer support. The arrangement also promotes communication between the faculties and administrations of the two programs.

Faculty members in the engineering program are available to students in the two-year engineering science program to give advice on career opportunities, job prospects, and local summer employment. In addition, faculty members in the engineering program teach some introductory courses in the two-year curriculum. Students in the two-year program are also invited to participate in student engineering societies based at the University of Kentucky.

Itasca Community College and University of North Dakota Two-Year Student Transfer Program to a Four-Year Engineering Program

Itasca Community College offers a unique engineering and physics program that includes classroom, laboratory, social, and residential areas in a 25,000-square-foot Learning and Living Center, aggressive recruiting efforts, intensive personal coaching to ensure retention and address students' educational and/or personal needs, block scheduling of classes, and project-based learning. From 1998 to 2002, 85 percent of Itasca's 140 students in engineering and physics transferred to four-year engineering programs, the majority at the University of North Dakota. From 1999 to 2003, 110 students graduated four-year programs with A.S. degrees from Itasca; 100 (91 percent) of them either became engineers or are on track to do so. The Itasca-University of North Dakota articulation partnership is one of the "second-level" partnerships described at the workshop—it is longitudinal (i.e., it covers middle school to university), has dual programming, and most important, both partners are fully committed to the program.

San Antonio Community College

Research by Dimitriu and O'Connor (2004), professors at San Antonio Community College, identified the following key elements for a successful partnership:

- mutual respect and trust between the participating institutions at all levels
- open, ongoing dialogue between faculties and administrators
- common objectives, strategies, and advertising
- coordinated programs and activities, including joint projects and cultural activities
- ongoing efforts to refine and align lower-level course offerings at the community college level
- visible presence of partners on each other's campuses
- sharing of facilities and faculty

Using these findings as a guide to developing effective transfer partnerships, San Antonio College established articulation agreements with engineering programs at University of Texas at San Antonio, Texas A&M University at Kingsville and Texas A&M University at Corpus Christi and has maintained its existing agreements with Texas A&M University at College Station, Texas State University in San Marcos, and University of the Incarnate Word in San Antonio.

The Regent's Engineering Transfer Program (RETP) at Georgia Institute of Technology

This program is designed to address a shortage of engineers in Georgia and increase access to engineering education for state residents. Five colleges (the term "community college" is not generally used in the Georgia educational system) and nine state universities participate in the program, which allows Georgia residents to begin their studies at any of the 14 institutions and then earn a bachelor of engineering degree at Georgia Tech (the only public engineering program in the state). Students accepted into the program complete their freshman and sophomore year courses at any participating college or university, and upon completion, transfer to Georgia Tech to earn selected engineering degrees.

This program offers many benefits to students:

- attendance at an educational institution close to home before transferring to the urban Georgia Tech campus in Atlanta
- lower costs for the first two years
- participation in cooperative programs (where they exist) near where they reside
- an invitation to Georgia Tech for a campus visit each spring to meet with engineering advisors
- a competitive advantage over non-RETP students transferring to Georgia Tech

The program also includes mentoring programs to support transfer students, collaborations between Georgia Tech faculty members and the 14 partner colleges, and partnerships with industry.

Prospective students must be Georgia residents, have a minimum mathematics SAT score of 560, have a minimum combined mathematics and verbal SAT score of 1090, and have a minimum high school GPA of 3.0. These standards are not so stringent as to discourage students but do increase the chance that students who enter the freshman year of the program will succeed. Students, who complete the two-year program, are guaranteed transfer, thus eliminating concerns about equivalency and the need for engineering faculty at Georgia Tech to evaluate students' eligibility on a course-by-course basis.

Benefits to the engineering program at Georgia Tech include increased diversity as transfer students from all over the state join the student body and continued top ranking of Georgia Tech among engineering colleges. In 2003–2004, the College of Engineering at Georgia Tech had 6,545 undergraduate and 3,298 graduate students and awarded 1,386 B.S. degrees, the most undergraduate degrees awarded by a U.S. college of engineering. In that same year, the engineering program was ranked first in the number of engineering degrees awarded to women and underrepresented minority students. Thus, RETP was identified by the committee as an exemplary approach to increasing diversity in engineering education, along with its exemplary approach to articulation and transfer.

Transfer Opportunity Program (TOP) at the University of California, Davis

The Transfer Opportunity Program (TOP) is a collaboration between the University of California, Davis (UCD), and 15 northern California community colleges. TOP coordinators from the Undergraduate Admissions Office at UCD regularly visit participating colleges to counsel students and parents on admission to UCD; preparation for majors and general education requirements; and financial aid, housing, internships, study abroad, and other student services. Engineering advisors also provide transcript evaluations, seminars on academic and career opportunities in engineering, and guided tours of the UCD campus.

UCD executes a formal written Transfer Admissions Agreement (TAA) for the prospective student specifying which courses must be completed and the required GPA. The TAA, written a year before transfer, lists the requirements for a chosen major and guarantees admission to UCD in that major if the requirements are met. The agreement is signed by the student, the community college counselor, and a UCD representative. College of Engineering TAA advisors provide ongoing advice by telephone and e-mail. California community college transfer applicants who have completed all the required lower-division coursework offered at the educational institution they attended are given the highest priority.

Support for transfer students as part of the TAA process includes orientation and counseling early in the summer before fall enrollment; full-time undergraduate staff advisors in each engineering department; mandatory annual counseling; early intervention for students in academic difficulty; and additional counseling by the Mathematics Engineering Science Achievement Program (MESA).

These support services are designed to help students overcome two common problems: (1) different lower-division requirements at University of California and California State University campuses; and (2) the change from the community college semester system to the UCD quarter system.

The graduation rate for transfer students, who make up 24 percent of engineering B.S. degree recipients at UCD, is the same as for students who begin their engineering studies at UCD—84 percent of each group earn B.S. degrees in engineering; another 5 percent pursue other majors or graduate from other colleges.

Evaluations of Transfer Programs

Many workshop participants pointed out a need for better evaluations of transfer programs. Evaluations should include: definitions of positive outcomes of diversity; assessments of learning outcomes; and assessments of shared-learning, outcomes-based objectives. The main concern was that articulation agreements may focus only on easily measured outcomes, such as the number of credits or courses, rather than more important learning outcomes and competences. In fact, there is a fundamental disagreement between two-year and four-year programs about what evaluations should measure. Another serious problem is that many educational institutions do not collect or analyze data on students that would support assessments; even when data are collected, this is not done uniformly.

One exception, however, is the California Council on Science and Technology (CCST) Critical Path Analysis of the California Community College System in the Generation of Science and Technology Graduates (CCST, 2002). The California Council on Science and Technology undertook a comprehensive evaluation of California's high-technology infrastructure for the purpose of revealing weaknesses in the science and technology educational system and using the results to develop rational responses to the changing technology environment. The study includes every component of the California educational system, from kindergarten through graduate school; it assesses the status of each component of the overall system, identifies strengths, weaknesses, and bottlenecks in the educational pipeline that prevent students from obtaining college degrees.

The CCST study includes six segments of the educational system, including community colleges and four-year institutions. The collection and analysis of statistics on the role of community colleges in the education of engineering, as well as science, technology, and mathematics graduates, is an exemplary approach that could serve as a model for other states. The study revealed a need for rigorous research to address the following issues:

- the dynamic flow of science, technology, engineering, and mathematics (STEM) students through the educational system
- reliable data, including the total number of students and the number of students in various categories
- points of entry (e.g., from high school or the workforce)
- outputs (e.g., time to degree, certificate, transfer, or cessation of study)
- factors that affect the flow of STEM students
- opportunities for success and barriers to the success of STEM students

Organizational Efforts to Enhance Communication

Washington Council for Engineering and Related Technical Education

This voluntary council, established in 1970 with the approval of the Washington State Legislature, comprises faculty and administrators from 17 community colleges with engineering transfer programs, four community colleges with engineering technology programs, five universities that offer programs leading to a B.S. in engineering, and three universities with programs leading to a B.S. in engineering technology.

The ultimate purpose of the council is to develop a statewide approach to engineering education and present a unified front to the legislature. The council's immediate goals include: improving communications among two-year and four-year institutions involved in engineering and engineering technology programs; coordinating engineering programs statewide; and developing personal contacts among faculty members. According to faculty members and administrators who have participated, the council is promoting a new statewide A.S. degree to ensure standardization, provides members of the state committee on the A.S. degree, and represents engineering education to the state government. The council meets twice a year in different locations throughout the state, alternating between two-year and four-year institutions. Leadership also alternates this way. The council is currently developing a Web site, *http://www.wcerte.org/*.

California Engineering Liaison Council

Established in 1947, the California Engineering Liaison Council is made up of administrators, faculty members, and staff members from colleges of engineering at University of California and California State University campuses, community colleges, and independent universities and colleges. The council meets twice a year, alternating between northern and southern California.

The council serves the following functions:

- exchanging information and developing recommendations for action in engineering education
- improving communications in matters of mutual interest
- developing, reviewing, and acting on proposals for policies, plans, and procedures that affect engineering education
- advising individuals and groups on promoting effective engineering education

The council is currently involved in activities to improve articulation between two-year and four-year programs; promote the uniformity of basic concepts in core courses; provide up-to-date, accurate guidance information; and collect and evaluate statistics on the number of incoming students and the number who complete engineering majors.

Two Year Engineering Science Association of New York State

The Two Year Engineering Science Association of New York State (TYESA) is an organization comprised of members from institutions of higher education in New York state that grant an A.S. degree in engineering science and adhere to curriculum guidelines set by the organization. Liaison membership is available to institutions and agencies interested in participating in TYESA meetings and other events.

TYESA member schools agree to adhere to the following curriculum, which was approved in 2000: four mathematics courses (calculus I and II, differential equations, and calculus III or linear algebra or statistics); one chemistry course; four courses in English and the humanities/social sciences; one computer programming course (mathematical software or programming); one introduction to engineering course; and a design experience—as a separate course or integrated existing course).

A minimum of six elective courses are required, at lease two of which must be in engineering science. Elective courses available to students include statics, dynamics, circuits and networks, strength of materials, thermodynamics, materials science, graphics, engineering laboratory, monoprocessor applications, organic chemistry I and II, modern physics, linear algebra, general biology I and II, chemistry II, digital electronics, and computer science I and II.

TYESA has six goals:

1. To provide a forum for the free expression of ideas and opinions of members on engineering education.
2. To maintain communication between two- and four-year colleges.
3. To provide a continuing review of the quality, content, and goals of the two-year engineering science curriculum.
4. To facilitate transfer from two-year to four-year colleges.
5. To provide a vehicle for joint action on all matters of mutual interest and benefit to engineering science programs offered by member institutions.
6. To maintain communication with the New York Department of Education and other regulatory agencies.

The State University of New York (SUNY) TYESA maintains a Web site that lists curriculum guidelines, upcoming events, links to member schools and four-year liaisons, and information about its annual design-and-build competitions.

The three programs described above have demonstrated innovative approaches to enhancing the transfer mission, but a number of issues are still unresolved. Research to identify areas for improvement in the transfer mission and to measure the effectiveness of current strategies and solutions would be useful.

CONCLUSION

The degree of institutionalization and regulation through oversight by a statewide entity varies among articulation agreements. Some of the agreements profiled at the workshop were developed because individual faculty members and administrators at a community college or four-year institution wanted to work together. These agreements are typically contingent on the personal relationships of individuals who are committed to the transfer mission. Other agreements derive from a top-down, statewide initiative to improve articulation and transfer of community college students. The committee noted three types of articulation agreements, based on the curricular scope of the agreement: course-by-course, departmental, and institutional.

Notable programs featured at the workshop had adopted a variety of approaches based on comprehensive articulation agreements, and they do appear to increase the likelihood that community college students will transfer to four-year engineering programs. Exemplary agreements include frequent interactions between (1) faculty members and administrators at community colleges and four-year institutions; (2) between two- and four-year partner institutions; (3) with high school (even middle school) students and teachers in surrounding communities; (4) and with parents, teachers, counselors, and leaders of youth organizations and educational outreach programs.

Communication between faculty members and administrators at partner institutions cited by workshop participants included campus visits, regularly scheduled local and statewide conferences and councils, and exchanges of course descriptions and curricular changes. Vehicles for communication with students and the people who advise them include print and Web advertisements; information on course and credit requirements, articulation agreements, and sources of advice and financial assistance; visits by two- and four-year faculty members to high schools and to one another's campuses; career fairs; and transfer counselors at both two- and four-year partner institutions.

Cooperation includes faculty member exchanges; a single registration form and fee for community college and transfer students; shared campuses or facilities; inclusion of community college students in engineering competitions and other events at four-year institutions and student engineering societies; access for transfer students to financial assistance, mentors, internships, and co-op opportunities; support for bridge programs and other outreach activities to interest more women and underrepresented minority students; and recruitment of high school students to community college A.S. degree programs.

Conclusion 2-1 Some articulation agreements are better than others. The most effective agreements provide for a seamless transfer of community college students to four-year engineering programs. These agreements are also characterized by continual improvements in response to changing engineering curricula at four-year educational institutions; institutionalized partnerships between community colleges and four-year institutions; and frequent interaction and collaboration between faculties and administrators of partner institutions.

Conclusion 2-2 The mission of effective transfer partnerships is "second-level articulation," That is, a focus on *transfer outcomes* rather than the *mechanics of articulation* (e.g., course credits, content and the exact sequencing of courses). The success of second-level articulation depends on communication and cooperation between transfer partners in pursuit of the same goal—the seamless transfer of community college engineering science students to four-year engineering programs and their attainment of a B.S. or advanced engineering degree.

Conclusion 2-3 Second-level articulation requires a culture in traditional engineering programs of focusing on the retention of engineering students, including transfer students, by providing a supportive educational environment.

Conclusion 2-4 Students who complete the A.S. degree before transferring are most likely to complete an engineering program and receive a bachelor's degree. Students who do not take enough engineering courses and transfer too soon often run into problems and are less likely to complete a degree. Data presented by a representative of Rochester Institute of Technology (RIT) at the workshop show that the persistence rate of transfer students is positively correlated with earning an A.S. degree prior to transferring.

Conclusion 2-5 Course-by-course articulation systems may discourage students from completing the A.S. degree for two reasons: (1) they see no benefit in completing a degree that includes courses that are not required at the four-year school and are, therefore, not accepted upon transfer; and (2) faculty members at the four-year college may tell them they do not need some courses to transfer. A block transfer agreement that gives premium transfer credits for completing the A.S. degree might encourage students to stay the course until graduation from two-year programs.

Conclusion 2-6 Uniformity between two-year engineering science curricula and lower-division courses at four-year engineering programs is desirable but not sufficient for seamless transfers of community college students. A minority of community college representatives at the workshop argued that articulation agreements should be more flexible—i.e., community colleges should not be expected to match their curricula exactly to four-year engineering programs. The committee concludes that greater flexibility, without compromising standards, could be achieved by ensuring that engineering pedagogy is less course driven and more outcomes based.

3

Recruitment and Retention

In this chapter, the committee discusses the recruitment and retention of students interested in pursuing a transfer pathway to a baccalaure ate or advanced degree in engineering. The chapter includes (1) descriptions of efforts by two-year institutions to attract and retain students in engineering science programs and prepare them to transfer to four-year engineering programs and (2) descriptions of efforts by four-year institutions to recruit students from community colleges and retain them in their engineering degree programs. The recruitment and retention of women and minority students, who are underrepresented in engineering, is also a focus of attention.

The National Science Board (2004) recently observed "a troubling decline in the number of U.S. citizens who are training to become scientists and engineers, whereas the number of jobs requiring science and engineering . . . training continues to grow." The board further observed:

> . . . if the trends identified in *Indicators* 2004 continue undeterred, three things will happen. The number of jobs in the U.S. economy that require science and engineering training will grow; the number of U.S. citizens prepared for those jobs will, at best, be level; and the availability of people from other countries who have science and engineering training will decline, either because of limits to entry imposed by U.S. national security restrictions or because of intense global competition for people with these skills.
>
> Even if action is taken today to change these trends, the reversal is 10 to 20 years away. The students entering the science and engineering workforce in 2004 with advanced degrees decided to take the necessary math-

> ematics courses to enable this career path when they were in middle school, up to 14 years ago. The students making that same decision in middle school today won't complete advanced training for science and engineering occupations until 2018 or 2020. If action is not taken now to change these trends, we could reach 2020 and find that the ability of U.S. research and educational institutions to regenerate has been damaged and that their preeminence has been lost to other areas of the world.

These statements underscore the importance of recruiting and retaining students who are U.S. citizens to the field of engineering.

KEY CHALLENGES

One of the critical challenges facing community colleges is increasing awareness of the opportunities they can offer engineering students. Students continue to enter community colleges without realizing they can obtain a four-year degree in engineering by beginning their studies at a community college and transferring to a four-year university engineering program. In addition, students and parents are often unaware of other benefits offered by community colleges, such as lower costs and flexible class scheduling. There is also a widespread belief that the education provided by community colleges is inferior to that of four-year institutions. Participants in the workshop also pointed out other challenges:

- inadequate or nonexistent guidance counseling in high schools
- a lack of advertising by community colleges and state agencies
- a failure of community colleges and four-year educational institutions to reach out to local high schools and to solicit the help of alumnae who could serve as role models and mentors
- the lack of data tracking outcomes for transfer students, which could demonstrate the viability of the community college pathway to engineering degrees
- a lack of recognition, guidance, assistance, and cooperation from four-year educational institution

To significantly increase the number of students who embark on the community college pathway to engineering, four-year schools will have to use their brand images to promote community college programs, perhaps by developing joint admission and recruitment programs with two-year schools. For example, a four-year school's name could be listed next to the engineering science major on the community college application, and promotional materials and high school outreach programs developed jointly should prominently feature the names of both schools.

Workshop participants generally agreed that strong partnerships between community colleges and four-year engineering programs improve student recruitment and retention for both institutions. Community colleges that reach out to potential students through a variety of messages and media and demonstrate that they have a proven record of success in preparing students to transfer to an engineering degree program are more likely to succeed in their recruitment and retention activities, especially if they have an established articulation agreement with a four-year institution. Four-year institutional partners also benefit by being able to draw on an expanded, and in some cases more diverse, recruitment pool that includes talented community college students. Moreover, four-year institutions will have better retention rates when they work together with community colleges to improve the preparation of students to pursue upper-division engineering courses.

Awareness of the benefits of a community college education varies among communities and states. According to workshop participants, community colleges frequently are perceived as a less beneficial alternative to four-year institutions for beginning postsecondary education suggesting that community colleges must "market" themselves better by building bridges with both high schools and four-year institutions to get their message across. Both two- and four-year institutions should make more use of the Internet, especially to disseminate transfer information on institutional websites.

Despite their recruitment efforts, community colleges report that many students who come to their campuses do not realize they can earn a four-year degree in engineering by starting on the community college pathway. This is especially true for students from inner-city schools and low-income communities, many of whom enter the educational pipeline through community colleges. Articulation agreements with four-year institutions are useful only if students and their parents are aware of them and are given proper guidance during the first two years of their coursework (Rifkin, 1998).

Proximity often plays an important role in students' decisions to attend community colleges. In many cases, four-year institutions with engineering programs are located in distant communities, and community colleges are nearby. Four-year institutions can improve their recruiting of students by making their campuses accessible to students who prefer not to travel long distances. Distance learning may be an option for such students, although it is not known how many community colleges offer distance learning options or how many students are involved. Some of the institutions represented at the workshop offer distance learning as a way of recruiting and retaining students.

Another challenge for both community colleges and four-year institutions is to expand collaborations with high schools. As a number of workshop participants noted, early competence in mathematics is key to successful matriculation in engineering. Another prerequisite is strong reading comprehension skills. Advanced high school students could also be encouraged to take concurrent courses at the community college level, which may improve their chances of completing an engineering degree. The committee and workshop participants agreed that encouraging students to earn community college credits in the twelfth grade was a good way to counteract "senioritis," which is characterized by a drop in interest in academics as students approach graduation.

Students in two-year programs often work part time or even full time, and many have family commitments. Working students frequently carry fewer credit hours and may become discouraged by how long it takes to complete the A.S. degree. Data collected by the National Center for Education Statistics suggest that many students who begin their higher educations at community colleges do not have adequate skills in mathematics and reading (Hoachlander et al., 2003). Students who require enrichment work to build up their skills before they can enroll in engineering science courses face higher costs and, typically, longer to earn their degrees. Some states require that community college students take courses that are not required of students at four-year institutions, for example, physical education, which may add to their financial burden and length of time before they are eligible to transfer.

The growing diversity of the U.S. population and the number of college students who are first-generation Americans means that transfer students have increasingly diverse racial, ethnic, and socioeconomic backgrounds, often quite different from those of other students and faculty members. Engineering faculty members and administrators must recognize these differences and take into account that they may affect not only the social integration of transfer students, but also their academic performance.

Tsapogas (2004) notes that GPAs tend to be lower for transfer students:

> Science and engineering graduates with lower undergraduate grade point averages (GPAs) were more likely to have attended community college than were graduates with higher grade point averages. Fifty percent of S&E [science and engineering] graduates with less than a 2.24 GPA (mostly C's) reported that they had attended community college before receiving their S&E degrees, compared with 42 percent of those with an undergraduate GPA of 3.75–4.00 (mostly A's).

However, in a meta-analysis of transfer shock—the temporary dip in the GPAs of transfer students in the first or second semester after transfer—Diaz (cited in Laanan, 2001) found that in most cases the drop was not dramatic and that many students recovered quickly. Although the GPAs of transfer students entering four-year engineering programs tend to be lower than those of four-year-only students (which translates to lower GPAs at graduation), transfer students are as likely, or more likely, than four-year-only students to meet academic expectations. Their retention was reported by workshop participants as equal to, or greater than, the retention of four-year-only students.

Adelman (1998) found that:

> among all students who reached the threshold of the engineering path and attended four-year colleges, the proportion of community college transfer students who completed bachelor's degrees in any field was almost indistinguishable from the proportion of students completing bachelor's degrees within four-year college attendance patterns, and the comparative proportions of these two groups completing degrees in engineering is not statistically significant.

Specifically, 65.8 percent of community college transfer students completed a bachelor's degree in engineering, compared with 60.4 percent of students in four-year-only institutions who persisted to the junior year.

In the literature (and corroborated in the workshop), the lack of financial assistance for community college students is cited as a major barrier to their retention, both before and after they transfer to four-year engineering programs. While most states increased their spending on community colleges in the 2003–2004 academic year, tuitions also increased—on average by 7 percent nationwide (Katsinas et al., 2004). In addition, because of different institutional contexts, the sources of financial aid may be different for the subset of students who eventually transfer to four-year programs than for community college students as a whole. The weight of testimony from workshop participants concerning this issue, led the committee to conclude that there is a pressing need for new programs, as well as expanded existing programs and sources of financial assistance, for community college students and transfer students.

Community college students often need financial assistance for many reasons. First, as noted above, many are working at least part time while they attend college. Financial assistance would enable them to focus more on their education. Second, students from the racial and ethnic groups that are underrepresented in engineering are more likely to come from low-income families for whom the cost of a community college education is a financial burden. Adelman (2004) suggests that low socioeconomic status lessens the likelihood of a student completing a degree. Zamani (2001) notes that low-income students and non-Asian minority students have lower transfer and program-completion rates.

A recent report by the Pell Institute (2004) observed, "Most students from low-income families never consider going to college, and those who do tend to go to community and for-profit colleges." The report cites data from the first national study to measure opportunities for low-income students to access and succeed in higher education in recent academic years (Table 3-1).

A 2004 study found that financially needy first-generation college students are much more likely to complete an associate degree if they attend an institution with reliable class schedules and an easy-to-navigate bureaucracy (Person & Rosenbaum, 2004). The authors observed that "College access does not always translate quickly or easily into college success. We need to look at how colleges' organizational environments might be more supportive for highly disadvantaged students." The study identified three key organizational structures that seemed to raise graduation rates:

- "One-stop shopping"—prospective students could often enroll, register, and apply for federal financial aid by working with a single person in a single afternoon. This structure was typically available in private, but not public two-year colleges.
- Predictable and streamlined curricula. As noted above, low-income students often face an array of demands from jobs and family members. They are more likely to complete an associate degree if they can be confident that their courses will be offered in a regular sequence at convenient times of the day. Students at the public colleges often reported that classes had been canceled at the last minute, or that some of their required courses had been offered at night, but others during the day.
- Low counselor-student ratios. Because the private colleges' accreditation partly depends on their graduation and job-placement rates, they closely monitor their students' progress.

The Pell report notes that the "stratification by students' income" seems to be increasing, that is, low-income students are increasingly at-

TABLE 3-1 Type of Institutions Students Attend, by Family Incomes

	Family Income (percent)		
Type of Institution	Under $25,000	$25,000–$74,999	More than $75,000
Community college	20	59	21
Private two-year institution	22	50	28
Public four-year institution	11	48	41
Private four-year institution	8	35	57

SOURCE: Pell Institute, 2004.

tending two-year and non-degree-granting institutions. The report concludes that, "If low-income students are unable to go to four-year colleges and obtain a bachelor's degree, their aspirations and achievements will be limited" (Pell Institute, 2004).

Transfer to a four-year engineering program almost certainly exacerbates the financial burden of low-income students and their families, not only because tuitions are higher than at community colleges, but also because demands on students' time increase and course schedules are less flexible. The rigor of upper-division engineering courses also limits the number of hours students can work.

Other students who enter engineering science programs better prepared academically and with more abundant financial resources may leave programs because the traditional curriculum—lacking design- and project-based content—fails to communicate what engineers actually do. This is also a problem in four-year engineering programs. Four-year engineering faculty could collaborate with two-year institutions to develop such courses.

Retention strategies by a number of community colleges alone and in collaboration with four-year partners include scholarships, internships, faculty advisors, mentoring by peers, study centers and mentoring programs, mathematics laboratories, weekend and evening hours for laboratories and study centers, engineering clubs, collaborations with local industry, and joint events with four-year partners to highlight the project and design elements of engineering.

EXEMPLARY APPROACHES TO RECRUITMENT AND RETENTION

Recruitment strategies by a number of community colleges represented at the workshop had many common features:

- visits to local high schools
- direct mailings to principles, counselors, and students who were not admitted to four-year institutions and others, whose names may be supplied by four-year institutions in the area
- advertisements via posters, brochures, and websites
- participation in joint recruitment events with four-year partners
- providing speakers for physics and other classes at local high schools
- inviting college representatives to speak at local career days
- sponsoring senior preview days in the spring
- inviting community college students to robotics competitions and other project or design events held on university campuses

- engineering camps and after-school programs
- providing scholarships and internships to transfer students
- recruitment events at local military bases
- recruitment events at technology days

Less frequently mentioned features included visits to middle schools, collaborations with K–12 teachers, and outreach to high school students through invitations to observe or compete alongside community college engineering students (e.g., in robotics and other design-and-fabrication competitions) in events held on university campuses.

North Carolina State University/ Lenoir Community College Partnership

North Carolina State University (NCSU) provides its transfer partner, Lenoir Community College, with the names of students who were not accepted into its freshman class. Direct mailings are then sent to these students to inform them of the possibility of beginning their engineering degree work at Lenoir and transferring to NCSU as juniors. Lenoir and NCSU have just begun a program to work with K–12 students in the region to stimulate interest in engineering.

University of Texas at San Antonio Pre-freshman Engineering Program

Founded in 1979, the University of Texas at San Antonio Pre-freshman Engineering Program (PREP) is designed to address the problem of recruitment and retention of students in engineering programs. The program has been so successful that it has been replicated at several locations in Texas and other states.

PREP is designed to motivate middle school students to begin studying for careers in science, engineering, and technology. According to the program website (*www.texprep.org*), 90 percent of students who participated in the program have graduated from college, and 52 percent of them majored in mathematics, science, or engineering. The most recent survey (in 2002) reported figures of 88 percent and 50 percent, respectively. Whether the small decline was part of a trend or simply the result of statistical fluctuation, faculty members felt that the rates could be improved.

Because students usually complete PREP by their first year of high school, there are two more years for reinforcing their confidence that they are capable of earning a bachelor's degree in a science, technology, engineering, or mathematics (STEM) field. To bridge the two-year gap, a new program was introduced at San Antonio College to provide a transition between the completion of PREP and the first year of college. This program—Early Development of General Engineering (EDGE)—is described below.

West Kentucky Community and Technical College and University of Kentucky

These two educational institutions have established a partnership for recruitment that includes collaborative scholarship programs and attendance at college nights at local high schools. They share advisory committee and board members and collaborate on infrastructure development (they share a common campus) and have a common tuition-billing program.

San Antonio College Early Development of General Engineering Program (EDGE)

The EDGE program was initiated during an eight-week summer session at the college in 2003. Twenty to twenty-five tenth- and eleventh-grade students were enrolled in two college courses: college algebra and introduction to engineering. The two classes met from 9 a.m. to noon, Monday through Friday. Afternoon activities consisted of supervised study (SS1) and student success sessions (SS2) from 1 p.m. to 4 p.m. In the SS1 sessions, groups of about 10 students worked together on homework and group projects, received assistance with assignments, and built a sense of community and shared success under the supervision of a leader or mentor. Key elements of the program were collaborative learning, peer support, workshops on study techniques, test taking, guest speakers, and special presentations on engineering. Four field trips introduced students to engineering in two private companies, one agency, and one university. Program results were compiled from the paper, "Getting an EDGE in Engineering Education" by O'Connor and Dimitriu (2004).

MESA MSP and MSTC Programs

MESA (mathematics, engineering, science achievement) operates two programs at the precollege level: the MESA School Program (MSP) and the MESA Success Through Collaboration (MSTC) Program. MSP assists students in middle schools (grades 6–8), high schools (typically grades 9–12), and some elementary schools to boost their performance in mathematics and science and become eligible to enroll in a college/university program in mathematics, engineering, or science. MSP offers individual academic plans, academic excellence workshops, training in study skills, day academies, career and college exploration (e.g., guest speakers and field trips to show students different college and career opportunities, including engineering), parent leadership development sessions, and teacher training opportunities.

The MSTC program (which has been eliminated in at least one state since the workshop was organized due to budget cuts) was designed to build partnerships among individual schools, American Indian communities, and American Indian education centers. Targeted at a population that has little presence in engineering, this program is designed to increase the visibility of engineering as a career choice and provided academic skills necessary to entering a college or university engineering program. A unique feature of MSTTC is that its sites were located in remote rural areas, introducing mathematics-based career options in these communities. MSTP offered the same academic enrichment components as MSP.

MESA Community College Program

MESA Community College Program (MCCP) provides mathematics, engineering, and science enrichment to community college students to enable them to excel academically and ultimately transfer to four-year institutions as mathematics-based majors. The program establishes academic community centers on campuses where most students are commuters and peer support and information sharing are scarce. The program is supported by industry, which hopes to help students learn firsthand about career options, scholarships, internships, and special programs.

MCCP offers a range of activities and services to community college students to improve retention rates and prepare students for transfer to four-year institutions. MCCP activities and services include: academic excellence workshops; an orientation course; assistance in the transfer process (help completing applications, counseling, and field trips to universities); a student study center on campus; and professional development workshops.

MCCP's career-advising services expose students to a variety of mathematics, engineering, and science career options through industry mentors, field trips, job shadowing, career fairs, and internships. The program links community college students with student and professional organizations through speakers' series and tours of companies. Corporate representatives, including many alumni, participate on the board and provide an important connection between students and companies; they also make their companies' resources available to MCCP students in the form of scholarships, strategic planning resources, summer internships, and field trips.

Prince George's Community College

The STEM Collegian Center, a new program at Prince George's Community College, provides both peer and faculty mentoring and advising to students in all STEM disciplines. Faculty members also work individually with students during the "Introduction to Engineering" course to help them become comfortable with seeking out career counseling and academic assistance from faculty members.

Three Rivers Community College (TRCC)

Every academic year, the director of the nuclear degree program takes all of the students in the Three Rivers Community College (TRCC) program on a tour of the nuclear engineering and health physics departments at the University of Massachusetts, Lowell, to acquaint them with the academics of these programs and inform them about internships and permanent job opportunities. During the tour, graduates of TRCC meet with students in the program to discuss their personal experiences.

Another feature of the program is that second-year students who are doing well in their coursework are asked to make themselves available as tutors for freshmen students. They set up mutually agreeable meeting times and assist the freshmen, as needed. A unique initiative at TRCC is a program in which designated third-year students act as tutors for second-year students. The third-year students are graduates of the nuclear engineering program who have been selected to stay on for a third year under full scholarship to obtain an additional A.S. degree in either electrical or mechanical technology.

In addition, the 24 members of the Nuclear Advisory Committee make themselves available to meet with students on a one-on-one basis to discuss career possibilities. Members of that committee are from the nuclear business community in Connecticut and department chairs from the four-year programs into which TRCC students normally transfer. The results of the TRCC program have been greatly improved by the exceptional financial support of the nuclear industry, a feature that cannot be replicated by other programs, although it may provide a model.

Jones County Junior College

A faculty member at Jones County Junior College (the only college represented at the workshop that continues to describe itself as a junior college) describes the Student Engineering Society as the best feature of the engineering science program and an important retention tool. The society is a student chapter of National Society of Professional Engineers (NSPE) sponsored by the local MESA chapter. The society sponsors monthly talks by individuals from various engineering disciplines and trips to a variety of industries where students can learn firsthand what engineers of various types do on the job.

Monroe Community College

The Monroe Community College (MCC) program provides students with design-and-build experience in several engineering science courses. In the first semester, students design the chassis, transmission, and other components of a small car powered with a DC motor. The students then fabricate the parts in one of two manual machine shops on campus, using mills, lathes, and other tools. At the end of the semester, the cars compete in sprinting, pulling, and climbing events.

In the second semester, students are assigned an electronic microcontroller design project. For example, students design and build alarm systems with infrared sensors, servomotors, switches, light-emitting diodes, speakers, and microcontrollers. In their final semester, students work in teams to design and build a working prototype to compete in the annual SUNY Engineering Science Association Competition. The top three MCC teams receive funding from the engineering club to travel to and compete at the SUNY Two Year Engineering Science Association (TYESA) Competition.

In the final semester of 2004, students designed and built robots that collected three red and three blue golf balls on an 8 ft × 8ft plywood surface and deposited them in the appropriate red or blue goal. These design-and-build experiences are extremely stimulating for students. Faculty members have observed that students work harder on projects than on other class assignments. Design-and-build projects also prepare students to be self-directed learners and to work effectively with a team of peers.

EXEMPLARY APPROACHES TO FINANCIAL ASSISTANCE

Overall, community college students have a decidedly lower socioeconomic status distribution than four-year students. A higher proportion of them (45 percent) are first-generation college students with English as a second language, and many come from Hispanic backgrounds. These characteristics correlate with attendance at K–12 schools in urban and impoverished areas. In addition, many of these students do not reach the threshold for majoring in engineering (precalculus, calculus, introduction to engineering design, and engineering graphics [CAD]). These students typically end up spending more time obtaining an A.S. degree than students who enter with the requisite STEM background. Many students work part- or full-time while they attend community colleges; thus, their enrollment in classes may be sporadic, and it may take them longer than usual to earn an A.S. degree or certification, especially if the courses they need are not offered at times when they can attend. Thus, due to their circumstances, students who are academically unprepared and/or financially disadvantaged are at risk for dropping out of community colleges before they earn an associate degree.

Three Rivers Community College

The Three Rivers Community College (TRCC) program is a technology degree program, rather than an engineering science program, that has partnerships with four-year institutions with engineering degree programs. Although the focus of this report is on the latter, some sophomore and junior students from engineering technology degree programs at community colleges and technical schools do enter four-year engineering programs.

The technology program is based on a unique partnership with a local nuclear facility. The Millstone Station nuclear facility offers as many as 17 scholarships each year to students in the TRCC nuclear program, providing tuition, fees, books, supplies, a monthly stipend during the academic year, a guaranteed 12-week summer internship, and access to job postings at the Millsone Station facility. Students who accept scholarships are under no obligation to work at the facility when they complete their studies. Graduates from the program can readily transfer into four-year degree programs in nuclear engineering nationwide as juniors. TRCC has articulation pacts with numerous four-year universities, all of which are accredited by Connecticut Engineering Accreditation Commission and Technology Accreditation Commission.

Merrimack College

Most of the students at Northern Essex Community College the partner institution of Merrimack College, are part-time students who work full or part time and cannot, therefore, attend Merrimack as full time students. To accommodate these students, Merrimack created an accredited, *part-time* electrical engineering degree for working students. Beginning in the fall of 2004, Merrimack also lowered its rates for part-time students from $760 per credit to $235 per credit. Merrimack College also offers science and engineering scholarships that enable community college students with a 3.5 GPA in an approved A.S. program to transfer to Merrimack with the same tuition and fees as for University of Massachusetts at Lowell.

CONCLUSION

The programs described above have developed innovative approaches to improve recruitment and retention, but a number of issues remain to be resolved. Further research might focus on identifying additional areas for improvement in recruitment and retention and measuring the effectiveness of current strategies.

Improving the community college pathway for students from underrepresented minorities and women will be important to diversifying the U.S. engineering workforce. To be successful, community colleges and four-year educational institutions will have to achieve their recruitment and retention goals for nontraditional students. Representatives of four-year engineering programs at the workshop reported that retention rates for transfer students was generally as high, or higher, than the rate for four-year-only students. However, the number of community college students with the potential to complete B.S. degrees in engineering who leave the engineering pathway prior to transfer is not known. Programs aimed at recruitment range in intensity from one-time activities (e.g., visits to local schools, summer camps, and career fairs) to comprehensive national programs that focus on both social integration of minority students and academic preparedness and excellence in STEM disciplines.

Broadly disseminated articulation agreements are essential to informing and reassuring students and parents that an engineering degree gained through the community college pathway is not only possible for them but is also highly probable if they follow the guidelines in the articulation agreements. In addition to articulation agreements, the recruitment and retention of engineering students requires close collaboration between transfer partners in a variety of other activities, including assistance with finding and applying to programs offering financial assistance.

A key challenge to recruitment and retention is the need for more public awareness (especially among first-generation and low-income students) that it is possible to earn a B.S. degree in engineering through the community college pathway.

Conclusion 3-1 Community colleges are in the best position to undertake outreach programs to K–12 teachers and students in their communities, using the successful communication and dissemination strategies described in this chapter of the report. Recruitment is most effective when data on the success of transfer students are available to potential students, parents, and high school teachers and advisors. Unfortunately, most two-year schools do not have the resources to compile and analyze these kinds of data.

Another recruitment challenge is the widespread perception that the first two years of undergraduate engineering education is characterized by a sink-or-swim culture, the goal of which is to weed out weak students (Seymour & Hewitt, 1997). Additionally, women are more likely than men to perceive that the culture of engineering schools is particularly inhospitable to women (Heyman, et al, 2002). Faculty and administrators of colleges of engineering must work with community college personnel to change the climate from intimidating to inviting and supportive. This will require that faculty members be aware of the social and cultural backgrounds and learning styles of women and students from underrepresented minorities (NAE, 2004).

The following activities can help create a welcoming environment, promote retention in A.S. programs, and increase the likelihood that community college students will transfer to four-year programs:

- joint participation by two- and four-year faculties in recruitment activities
- campus visits, faculty exchanges, and sharing of laboratories and other facilities
- academic counseling and mentoring of community college students by upper-division engineering students
- "leveling the playing field" between transfer and four-year-only students in subtle and not-so-subtle ways (e.g., GPA requirements)
- inclusion of community college students in engineering society chapters, build-and-design competitions, internships, and cooperative activities.

Conclusion 3-2 The lack of financial assistance from institutional, state, and federal sources is an enormous barrier to the recruitment and retention of engineering science students in community colleges. Additional financial assistance is also needed for students who transfer to four-year engineering programs to ensure that they can afford to stay in school until graduation. Funding could come from institutions of higher education or through fellowships and scholarships for transfer students funded by industry, and the federal government.

Research suggests that outreach to K–12 schools can encourage students to pursue engineering in college. The middle school years are a critical time for encouraging students, especially girls, to take the mathematics and science courses necessary to reach the threshold required to pursue an undergraduate degree in engineering (Adelman, 1999).

Often, only a small number of students at an individual community college are interested in engineering, which makes it difficult to justify maintaining a faculty that has engineering expertise. In these cases, it is also difficult to attract students to the program. A series of seminars about engineering and technology/career opportunities, supported by the faculty and graduate students of the four-year partner and engineers in industry from the surrounding community, would help administrators gauge the interest of local students in engineering and enable them to plan and advocate for more engineering science faculty, if appropriate.

Community colleges might also consider recruiting industry experts to teach on their campuses. Many practicing and retired engineers could contribute to engineering education and strengthen the links between the engineering curriculum and the real-world applications of coursework. After all, engineers *do* things.

Four-year educational institutions, especially Research 1 universities, frequently have partnerships with industry on research projects and sometimes for instruction. Community colleges do not have the same opportunities to work with engineers in research, but they might still attract engineers in industry who are interested in connecting with students, through teaching, especially hands-on, project-based coursework.

Conclusion 3-3 Community colleges could develop partnerships with industries in their areas to recruit interested and qualified industry engineers to demonstrate the practical applications of mathematics and introductory engineering coursework.

4

Curricular Content, Quality, and Standards

Engineering continues to change in response to new challenges and new technologies. Recent technological breakthroughs have been made in biotechnology, nanotechnology, information and communications technology, materials science, and photonics, and other fields. In response to changes in engineering, engineering education is also changing. As engineering becomes increasingly specialized, more material must be covered in lower division courses, as well as in junior and senior year courses. Fundamental changes have been made in engineering education in four-year B.S. programs: more project-based learning; the introduction of principles of design and other professional engineering features in lower division courses; and more emphasis on life sciences, interdisciplinary material, and liberal arts.

The increase in required prerequisite knowledge in related disciplines may increase the amount of coursework required for an A.S. degree at two-year educational institutions. In some states, community college students are already required to complete not only general education courses, but also courses in seemingly unrelated fields, such as physical education. As a result, the number of credits required for an engineering degree has increased. Research shows that students who earned a baccalaureate degree in engineering who started at four-year institutions completed an average of 149 credits; students who started at community colleges completed an average of 160 credits (Adelman, 2004).

There is a growing consensus among educators and policy experts that engineering curricula and pedagogy must be changed. Wulf and

Fisher (2002) have argued that engineering educational institutions are becoming increasingly out of touch with the practice of engineering:

> Not only are they unattractive to many students in the first place, but even among those who do enroll there is considerable disenchantment and a high dropout rate (of over 40 percent). Moreover, many of the students who make it to graduation enter the workforce ill-equipped for the complex interactions, across many disciplines, of real-world engineered systems.

A number of researchers have focused on the need to include design and build projects in lower division courses and to encourage research in the undergraduate curriculum.(Beston, 2004; Grimson, 2002; Seymour and Hewitt, 1997). The recommendations for how engineering education needs to change contained in the NAE 2004 report, *The Engineer of 2020: Visions of Engineering in the New Century*, are wide-ranging:

> Almost all discussion of educating the engineer of 2020 presumes additions to the curriculum—more on communications, more of the social sciences, more on business and economics, more cross-cultural studies, more on nano-, bio-, and information technologies, more on the fundamentals behind these increasingly central technologies, and so forth (NAE, 2004).

Changes have also been made in accreditation criteria, where the emphasis has shifted from student inputs to student learning outcomes. Criteria 3, Program Outcomes and Assessment, of ABET's Engineering Criteria states (ABET, 2004):

> Although institutions may use different terminology, for purposes of Criterion 3, program outcomes are intended to be statements that describe what students are expected to know or be able to do by the time of graduation from the program.

Engineering programs must demonstrate that their graduates have:

a. the ability to apply knowledge of mathematics, science, and engineering
b. the ability to design and conduct experiments, as well as to analyze and interpret data
c. the ability to design a system, component, or process to meet desired needs
d. the ability to function on multi-disciplinary teams
e. the ability to identify, formulate, and solve engineering problems
f. understanding of professional and ethical responsibility
g. the ability to communicate effectively
h. a broad education necessary to understand the impact of engineering solutions in a global and societal context

i. recognition of the need for, and an ability to engage in life-long learning

j. knowledge of contemporary issues

k. the ability to use the techniques, skills, and modern engineering tools necessary for engineering practice.

ABET mandates that each program have an assessment process and documented results. Programs must also show that the results are used to continue the development and improvement of the program. The assessment process must demonstrate that the outcomes of the program, including those listed above, are being measured.

KEY CHALLENGES

The discussion that follows addresses three questions raised in workshop presentations and committee discussions:

1. What material should be covered in A.S. programs at community colleges? Should community colleges strive to provide course content that is as close as possible to lower-division courses in four-year engineering programs?
2. What resources do A.S. degree programs need to prepare students to transfer and succeed in four-year engineering programs?
3. How can quality in A.S. programs be assured?

The first question relates to the development of a standard, lower-division engineering curriculum. In this regard, community college faculty and administrators are "shooting at a moving target." Courses that used to be taught at higher levels are increasingly being taught in lower levels of B.S. degree programs. Because of the lack of critical mass of students in community college programs and a lack of resources to recruit faculty with specialized expertise, this change presents serious problems for community colleges. In addition, institutions disagree about which courses are "lower level" and which are "upper level."

Developing a curriculum requires defining students' competencies in engineering and relevant cognate subjects after two years of college-level coursework. Although this was not included in the committee's charge, some obvious competencies can be listed:

Level 1: calculus sequence, physics, inorganic chemistry, and introduction to engineering (including design)

Level 2: calculus sequence, physics, introduction to engineering, statics, dynamics, fluids, thermo, circuits I and II, digital logic, mechanics, materials, organic chemistry, perhaps introduction to process design

Level 3: same as Level 2, but with a strong engineering-design and fabrication component in each course and with calculus and physics taught as engineering classes, perhaps at least partly by engineering faculty

How much variation could be allowed without compromising the principle of a common curriculum? For example, some universities insist that all engineering students must take their class in thermodynamics, which is designed to meet the specific needs of their mechanical engineering program. When elements of upper level courses are taught in the freshman or sophomore year, partnerships with community colleges can be threatened. A good measure of a successful partnership might be the willingness of a community college to commit to teaching the aforementioned classes at a level of expertise acceptable to the four-year partner as long as the four-year institution is willing to eliminate major-specific freshman and sophomore classes as a requirement for transfer. Another solution would be for the university partner to offer major-specific courses online to the community college(s) with which they partner.

Several workshop participants described how their institutions had used distance education to make courses more accessible to their students. For example, tribal colleges, which are predominantly two-year colleges, use distance learning to reach geographically isolated and dispersed groups of students. Distance learning may also appeal to some students whose work schedules prevent them from participating in classroom learning. In addition, distance education can also introduce community college students to four-year university faculty members and coursework; even laboratory work can be conducted via distance learning.

Research that has been done on the effectiveness of distance learning, is scant and inconclusive (Rovai, 2002). Two-year students typically choose community colleges for reasons that seem antithetical to distance learning, such as the personalized teaching/learning provided by smaller classes and more interaction with faculty—in other words, a sense of community. Given the limited amount of data available on the efficacy of distance learning for community colleges, the committee believes more research should be done in this area.

LOWER LEVEL CURRICULUM

Most of the workshop participants from two-year educational institutions pointed out the problems created for engineering science programs by the evolution of four-year curricula. As engineering curricula become more specialized, driven partly by the continual improvement process required by ABET 2000 criteria, community colleges are finding it increasingly difficult to offer courses or curricula that satisfy their four-year engineering partners. Because of capacity issues, many community colleges can only offer a single engineering sciences curriculum. As new courses are added to four-year engineering programs, the number of four-year engineering programs to which their students can transfer without losing credits is decreasing.

One of the weaknesses of articulation agreements is that changes made to four-year program curricula are often not reflected in the agreements. Some community college representatives also noted that they were unable to offer the number of lower-division courses defined by the articulation agreement because of insufficient faculty, laboratory facilities, and other resources.

Another weakness cited by representatives of community colleges was the failure of four-year institutions to consult with them about impending changes or to give them adequate time to adjust their curricula to reflect these changes. To ensure that community college faculty members and administrators have time to plan a response to changes, they must have timely, frequent communication with institutional partners.

Lack of communication between two-year and four-year educational institutions is a problem, especially for community colleges with numerous four-year transfer partners and those located a good distance from many of their partners. Since four-year institutions are in the dominant position in transfer partnerships, they must provide mechanisms for ensuring frequent communication with their community college partners to assist them in responding to curricular changes.

There is a growing consensus among engineering educators that the amount of application coursework should be expanded. Reducing the emphasis on strictly theoretical expositions and increasing the practical applications of engineering will benefit students in both two- and four-year institutions. Many presenters at the workshop cited a need for more project and design work in lower-division engineering courses for several reasons: to improve students' understanding of the relevance of coursework to what engineers actually do; to integrate more new technologies into coursework; to increase the emphasis on teamwork, communication skills, and skills required to operate in a global business environment; and, particularly for community colleges, to establish partnerships with local industries.

Several exemplary approaches to creating a more active learning environment were profiled at the workshop (the programs at Three Rivers Community College, Merrimack College, and Monroe Community College are described in previous chapters). Washington State University offers students a variety of real-world engineering (or architecture) experiences. Students can participate in a regional or national robotics competition, help create design-and-build projects for a concrete canoe, participate in bridge building, participate in designing a chemically powered car, or participate in environmental design competitions. Student groups have also built a solar-powered boat and a real airplane. More than 20 student clubs offer a wide variety of activities, and the graduate program in engineering offers international exchanges.

RESOURCES OF COMMUNITY COLLEGES

Two areas of concern for community colleges attempting to maintain high standards are faculty professional development and adequate infrastructure. Community colleges are teaching, not research, institutions. The committee is not able to generalize about these two areas of concern from the anecdotal information provided by workshop participants. The adequacy of facilities, in particular, is dependent on a variety of contextual factors the assessment of which is outside of the committee's charge. Testimony from workshop participants suggests that closer collaboration between four-year engineering programs and their two-year transfer partners—e.g., sharing facilities and faculty exchanges—would potentially enhance both the opportunities for professional development of community college faculty and the facilities available to community college students and faculty.

ACCREDITATION AND EVALUATION

It would be surprising if the subject of ABET accreditation of community college engineering science programs did not arise in the course of the workshop and the committee's deliberations, especially with regards to discussions of curriculum, standards, and quality. Currently, ABET does not accredit two-year engineering science programs; it does accredit two-year engineering technology programs. Community colleges are accredited by regional organizations. Some workshop participants expressed the view that an ABET-style accreditation process would not work for community colleges. Others felt that community college associations and engineering societies should address this issue. The committee's charge does not include addressing the question of whether or not engineering science programs should be accredited by ABET. However, further assessment of the value of accreditation, how accreditation might be best accomplished, and who might be best able to do so is warranted.

CONCLUSION

The central topic of this chapter is what a student pursuing a baccalaureate degree in engineering needs to know at the end of the first two years of study and how that knowledge can be demonstrated to a four-year institution and others. This question involves issues related to curriculum, pedagogy, and quality.

Conclusion 4-1 Institutions of higher education are addressing issues related to curriculum, pedagogy, and quality, but must do much more to resolve them.

Conclusion 4-2 As the trends in engineering education move toward greater diversity and specialization in the lower division course offerings of four-year engineering programs, engineering science curricula are less likely to cover the same material or achieve the same results. Thus, the need for communication and resource sharing between transfer partners and for the timely updating of articulation agreements is becoming more urgent.

Conclusion 4-3 The engineering education community, and the profession as a whole, would benefit from a discussion of the feasibility and desirability of standardized accreditation for community college engineering science programs.

Conclusion 4-4 More emphasis in the K–12 curriculum in U.S. schools needs to be placed on mathematics. Mathematics courses in engineering should put more emphasis on applied engineering examples.

5

Diversity in the Engineering Workforce

The engineering profession in the United States faces a diversity imperative. Historically, engineering has been a white male-domi nated profession. While other countries have been diversifying their engineering workforces, mostly in terms of increasing the number of women engineers, the "face" of the U.S. engineering workforce still does not reflect the diversity of the population. In spite of efforts by many individuals and organizations over the past 30 years, economic, cultural, educational, and institutional barriers continue to discourage or prevent women and underrepresented minorities from pursuing engineering education and the rewarding careers that can follow. To meet the need for the best and brightest individuals in engineering, as well as in science, technology, and mathematics, we must develop a domestic talent pool that looks very different from the one we draw on today.

KEY CHALLENGES

Since World War II, the United States has become increasingly dependent on talented individuals from abroad to meet its engineering needs instead of developing a broad domestic talent base that includes women and underrepresented minorities. Although the influx of foreign-born students into U.S. engineering schools and the engineering workforce has given U.S. engineering an international cultural and linguistic "feel," the creativity and ideas of American women and minorities have yet to be exploited.

In 2004, 46 percent of master's degrees and 57 percent of doctoral degrees in engineering were awarded to foreign nationals (EWC, 2004). However, the United States cannot continue to rely on talent from elsewhere to meet its engineering needs. As Chubin et al. (2005) and many, many others have observed, recent events have raised questions about the decades-long strategy of drawing engineering talent from other countries. We must begin to ask ourselves who should be admitted to U.S. universities, if we want graduates to stay here or return home, if national security concerns take precedence over the development of a global workforce, and if a profession rooted in American markets can thrive on increasingly foreign-born talent.

The demographics of the engineering workforce clearly reflect the problem. About one-third of the school-age population in the United States consists of underrepresented minority students. Women constitute more than half of the U.S. population and 60 percent of the total workforce. Jackson (2002) observed that women and minorities in the U.S. no longer are the underrepresented *minority*, they are the underrepresented *majority*. Nevertheless, the percentage of freshman women enrolled in engineering has declined recently, from a high of 19.9 percent in 1996 to 16.34 percent in 2004 (EWC, 2004).

The lack of diversity in the engineering workforce and in the engineering-education pipeline, poses significant, and growing, costs and risks for the engineering profession. First and foremost, the extreme underrepresentation of major segments of American society in engineering poses a moral and social dilemma, and, unless actions are taken to change the situation, the opportunity costs to the engineering enterprise and the nation will increase in the coming decades. Second, experience in industry and the classroom shows that creativity is increased and the range of potential solutions to problems is expanded when teams of people approach problems from diverse personal, cultural, and disciplinary perspectives. The scarcity of women and underrepresented minorities in U.S. engineering classrooms, research laboratories, design studios, and corporate boardrooms limits the perspectives and diversity of ideas/solutions (Wulf, 2002). Finally, although the long-term demand for engineering is notoriously difficult to predict, demographic trends guarantee that the current "underrepresented majority" in the United States will account for an increasing share of the population and workforce as the new century progresses.

Community colleges have long been recognized as providing opportunities to advance the goal of diversifying the U.S. engineering workforce, especially racial and ethnic diversity. Although the makeup of community colleges student bodies varies by geographic location, a larger percentage of students from some minority groups, notably Hispanics and American Indians, attend community colleges than white students. Concomitantly, community colleges have become an educational pipeline for underrepresented minorities entering the higher education system.

In 1992, enrollments of minority students accounted for about 25 percent of total enrollment in community colleges: 9.9 percent African American students; 1.1 percent American Indian students; 5.0 percent Asian/Pacific Islanders; and 9.3 percent Hispanic students (Table 5-1). By 1995, Hispanics enrolled in community colleges outnumbered African Americans (AACC, 2000). By 2002, the minority subtotal of community college enrollments had grown to 33 percent. Changes in the representation of subgroups of minority students between 1992 and 2002, although small, are likely to be predictive of future demographic changes. In 2002, African American students comprised 36.0 percent of minority enrollments—a slight decrease from 1992. Hispanic student representation rose to 41.2 percent, and American Indian student representation rose to 3.7. The largest increase—from 5.0 to 19.0 percent—was for Asian/Pacific Islanders (Phillippe and Gonzalez Sullivan, 2005).

Community college students differ from their counterparts at four-year institutions both demographically (i.e., in age, enrollment status, socioeconomic status, and educational background) and in student outcomes (i.e., grade point average, time to degree, and dropout rates). As Table 5-1 shows, community college students are more likely to be part-time students, more likely to be older, less likely to be dependents, more likely to have dependents, more likely to be minority students, and more likely to be working full time. In addition, some community college students are first-generation students, whose circumstances and worldview differ in many ways from students whose parents attended college (Inman and Mayes, 1999).

TABLE 5-1 Percentage of Bachelor's, Master's, and Doctoral Degrees in Academic Year 2003–2004, by Race/Ethnicity and Foreign-National Status

	Race/Ethnicity						
Type of Degree	White Male	White Female	Asian American	Hispanic American	African American	Native American	Foreign-National Status
Bachelor's	57%	12%	13%	6%	5%	<1%	7%
Master's	32	8	9	3	3	<1	46
Doctoral	27	7	6	2	2	<1	56

SOURCE: EWC, 2004.

TABLE 5-2 Percentage Distribution for 1999–2000 of Undergraduates According to Selected Student Characteristics, by Type of Institution.

Student characteristics	Total[1]	4-year total[2]	Public 4-year		Private not-for-profit		Public 2-year
			Doctoral	Nondoctoral	Doctoral	Nondoctoral	
Total	100.0	100.0	100.0	100.0	100.0	100.0	100.0
Sex							
Male	43.7	44.7	47.1	42.5	46.0	41.4	43.7
Female	56.3	55.3	52.9	57.5	54.0	58.6	56.3
Race/ethnicity[3]							
American Indian	1.0	0.9	0.7	0.8	0.5	1.2	1.3
Asian/Pacific Islander	6.4	6.7	8.0	5.7	8.8	3.4	6.1
Black	12.6	10.9	10.2	12.7	10.8	10.0	13.2
White	67.8	71.0	72.4	68.9	69.1	71.7	66.4
Hispanic	12.2	10.7	8.6	11.9	10.8	13.7	12.9
Age							
18 and under	9.5	10.9	10.8	10.0	14.0	10.1	8.5
19–23	47.7	58.3	62.1	53.3	65.2	51.8	36.6
24–29	17.0	14.5	15.5	16.7	9.9	13.1	18.4
30–39	13.9	9.4	6.7	12.0	6.1	14.1	18.5
40 and above	11.9	6.9	4.9	8.0	5.7	10.9	17.9
Average age	26.4	24.2	23.4	25.0	23.0	25.7	28.9
Dependency status							
Dependent	49.1	62.7	66.0	56.3	73.4	56.7	36.3
Independent	50.9	37.3	34.0	43.7	26.6	43.3	63.7
Respondent has dependents							
No	73.1	82.1	85.3	77.5	87.3	77.3	65.5
Yes	26.9	18.0	14.7	22.5	12.8	22.7	34.5
Single parent[4]							
No	86.7	91.0	91.9	88.9	93.1	90.2	83.6
Yes	13.3	9.0	8.1	11.1	6.9	9.8	16.4
Employment[5]							
Full-time	39.3	26.4	21.7	32.1	20.3	33.8	53.8
Part-time	40.8	50.7	53.9	47.7	52.4	46.3	30.4
Not working	19.9	22.9	24.4	20.3	27.3	19.9	15.8
Disability status[6]							
No disability	90.7	92.3	92.5	91.8	93.9	91.6	89.3
Disability or difficulty	9.3	7.7	7.5	8.2	6.1	8.5	10.7
Citizenship							
Citizens	93.0	93.9	93.8	94.0	91.7	95.5	92.2
Student and parent(s) foreign-born	4.2	3.5	3.3	4.1	4.5	2.4	5.0
Only parent(s) foreign-born	10.1	9.9	10.4	8.9	14.5	6.8	10.0
All other citizens	78.6	80.5	80.0	81.0	72.8	86.3	77.3
Permanent residents	5.1	3.9	4.1	4.6	3.7	2.5	6.1
Foreign students with visa	2.0	2.3	2.2	1.5	4.6	2.1	1.7
Home language							
English	87.3	89.3	89.1	89.1	87.2	91.4	85.6
Other than English	12.7	10.7	10.9	11.0	12.8	8.6	14.4
Attendance							
Full-time	52.1	68.5	69.1	62.7	77.3	69.3	30.5
Part-time	47.9	31.5	30.9	37.3	22.7	30.7	69.5
Delayed enrollment[7]							
Did not delay	54.5	67.8	71.9	62.7	72.5	61.8	41.3
Delayed	45.5	32.2	28.1	37.3	27.5	38.3	58.7
High school attainment[8]							
High school diploma	93.4	97.3	98.4	96.1	98.4	95.6	90.0
GED or other equivalency	5.2	2.1	1.1	3.4	1.0	3.7	7.9
High school completion certificate	0.3	0.2	0.2	0.2	0.2	0.3	0.4
No diploma or equivalent	1.1	0.3	0.3	0.4	0.4	0.4	1.7

See notes at end of table.

[1]Total includes students in institution types not listed here and students who attended more than one institution.

[2]Four-year total does not include students enrolled in private for-profit institutions.

[3]Students who identified their race as "other" (about 1 percent) are not included in this variable in order to make it comparable to the NPSAS 1993 categories. In 1999–2000, 2 percent of students chose more than one race. These students were then asked which single race best described them and were coded as such. American Indian includes Alaska Native, Pacific Islander includes Native Hawaiian, Black includes African American, and Hispanic includes Latino. Race categories exclude Hispanic origin unless specified.

[4]Includes some students with dependents other than children.

[5]Students who were employed full time worked 35 or more hours per a week.

[6]Students reported sensory or mobility limitation or another condition that caused difficulties.

[7]Students are considered to have delayed enrollment if there was 1 year or more between their high school graduation date and date of first enrollment in postsecondary education.

[8]Students who attended a foreign high school (about 1 percent) are not included in this variable in order to make it comparable to the NPSAS 1990 categories.

NOTE: Detail may not sum to totals because of rounding. See supplemental note 7 for definitions of the race/ethnic categories. See supplemental note 8 for information about postsecondary institution classifications.

SOURCE: U.S. Department of Education, NCES, 1999–2000 National Postsecondary Student Aid Study (NPSAS:2000).

Source: Department of Education, 2003.

EXEMPLARY APPROACHES

Several of the community colleges represented at the workshop had substantial numbers of minority students in their engineering programs. They attributed the numbers to the geographic area from which they draw students, rather than to strategies for recruitment and retention. A representative of one institution with a considerable number of black students noted that they were mostly from outside the United States and that, despite its efforts, little progress had been made in recruiting American-born black students.

National Action Council for Minorities in Engineering

The mission of the National Action Council for Minorities in Engineering (NACME) is to provide leadership and support for a national effort to increase the representation of African American, American Indian, and Latino women and men in engineering and technology and mathematics- and science-based careers. NACME is the nation's largest source of private-sector scholarships for minorities in engineering. Approximately 75 students annually are awarded full tuition and housing to attend one of nine institutions that participate in the NACME Engineering Vanguard Program. More than 15 percent of all minority graduates in engineering since 1974 received NACME scholarships (SWE, 2004).

Currently enrolled, full-time engineering students can also apply to participate in the NACME Corporate Internship Program, attend leadership-development seminars, and are mentored by corporate professionals. NACME also provides diversity interventions to schools, government agencies, organizations, and corporations. These include cultural audits, awareness seminars, and mentor training. For K–12 teachers, NACME offers professional development with training in use of the WorldCom MarcoPolo Internet Content for the Classroom, which provides standards-based materials in mathematics, science, and other subjects. NACME works with university admissions personnel and program directors to help them assess student potentials and trains them to use the assessment protocol developed in its Engineering Vanguard Program.

NACME hosts two Web sites—*www.guidemenacme.org* and *www.mathispower.org*—designed for precollege students, parents, and educators. *GuideMe* offers information on a variety of scholarship opportunities; pointers on preparing for and applying to engineering schools; and guidelines for selecting schools that meet student needs. The *Math Is Power* Web site features interactive games and puzzles, links to national Web-based resources, and a registry of local events.

The annual NACME Forum attracts public- and private-sector leaders from more than 200 organizations to discuss issues that effect minority achievement in engineering, share best practices, and develop strategies for recruiting and retaining minorities in engineering and technology.

Georgia Institute of Technology College of Engineering

Georgia Tech ranks first in the country in the number of engineering degrees it awards to women and underrepresented minorities. To attain that status, the Regents' Engineering Transfer Program was created to address the engineering needs of Georgia; to expand access to engineering education; and to do these two things in the context of the University System of Georgia. The program addresses the needs of many students: a desire to attend college close to home; a need to reduce the costs of education; and smaller classes, individual attention, and greater access to faculty members, typically offered by community colleges, to increase the chances of academic success. Georgia Tech's program is discussed in greater detail in Chapter 2 of the report.

Emerson Electric Company Minority Engineering Scholarship

This program, funded by Emerson Electric Company, provides opportunities for aspiring engineers to earn engineering degrees through St. Louis Community College and the University of Missouri at Rolla. Founded in St. Louis in 1890, Emerson is a global manufacturer of a variety of electrical and electronic products and systems for commercial, industrial, and consumer markets. The company has more than 100,000 employees at more than 300 manufacturing facilities worldwide.

The scholarship program makes it possible for participating students who keep their grades up to earn degrees in mechanical, electrical, or computer engineering from prestigious engineering schools—with nearly all expenses paid. Participating students follow a specific pre-engineering curriculum at one of the three campuses of St. Louis Community College and then transfer to the University of Missouri at Rolla to complete their engineering degrees. The Emerson scholarships—eight are awarded annually—cover the cost of tuition and fees for as many as six consecutive semesters at any of the community college campuses. The scholarship, which is awarded and reviewed on a semester-to-semester basis, continues when students transfer to the University of Missouri at Rolla.

CONCLUSION

Diversity among engineering students is an important goal, both for the U.S. S&E business community and for the higher education community. Equality of opportunity is a moral imperative in America, which is backed up by substantial legislation (e.g., the Civil Rights Act for minorities and Title IX for women). In addition, diversity has been shown to enhance education for all students (ACE and AAUP, 2000), and a diverse student body will lead to a more diverse workforce. Women and underrepresented minorities represent an untapped potential for the U.S. S&E workforce. Successful women and minority engineers in the workforce and on engineering faculties will also provide role models, thus reinforcing the confidence of others to follow in their footsteps.

Although some community colleges have very diverse student bodies, for geographic or other reasons, very few underrepresented minority students actually obtain engineering degrees. And four-year universities, although they are generally interested in increasing the diversity of their student populations, do not reach out enough to community colleges, by providing advisors, academic support, scholarships, student networking, and so on, to establish an effective mechanism for transfer.

The committee reviewed materials and heard expert testimony on the effectiveness of minority bridge programs in moving underrepresented minorities into engineering programs, as well as preparing them for the transition from high school to college (both two- and four-year programs) and from two-year to four-year programs. Some of these bridge programs begin working with students in middle, or even elementary, schools, and some focus primarily on easing the transition between community college engineering science or technology programs and four-year programs for all students. The MESA program, described in detail in Chapter 2 of the report was identified by the committee as an exemplary bridge program.

Conclusion 5-1 Lessons learned from programs designed for underrepresented groups in engineering could be applied to engineering students generally. This will require more research to identify the components of successful outreach/bridge programs that could be scaled to a larger engineering student population.

Conclusion 5-2 Four-year programs could be more proactive in developing bridge strategies with community college partners and collaborating in outreach to K–12 teachers and students. Four-year engineering programs could identify community colleges with large numbers of underrepresented minority students and establish relationships with these schools through faculty exchanges, invitations to students to visit their campus, speakers from their school, scholarships earmarked for transfer students, and other mechanisms that demonstrate their interest in attracting minority students to engineering careers.

Conclusion 5-3 Organizations in the engineering educational and professional communities could work together to increase awareness of the need for diversity in the engineering workforce and educate state and federal legislators. State and federal funding for community college students and incentives for four-year engineering institutions to reach out to community colleges and their students could help increase the number of underrepresented minorities in engineering.

Conclusion 5-4 Research could focus on identifying factors that are associated with the successful recruitment and retention of underrepresented minorities and women in science and engineering programs in two-year and four-year institutions and on assessing the success rate of the transfer mission for these students.

6

Data Collection

We know that 20 percent of engineers began their academic careers with at least 10 credits from a community college and that 40 percent of the recipients of engineering bachelor's and master's degrees in 1999 and 2000 attended a community college. However, there are a great many things about the community college pathway to engineering degrees and careers we do not know. A first step is simply to locate lower division engineering programs at community colleges—programs which may not be listed by a standard name. The committee's perception—echoed by the workshop participants—is that there are important gaps in the data, mostly in reference to students. For example, did students earn their credits from community colleges during their senior year of high school, during the summer before they entered four-year engineering programs, or as students in engineering science who left before obtaining A.S. degrees? This question and a number of others may not be answerable with existing data. What percentage of those who hold engineering degrees transferred to four-year engineering programs (1) with A.S.degrees and (2) without A.S.degrees? What percentage of students who transferred to four-year engineering programs went on to earn undergraduate or graduate degrees in engineering?

Answers to these questions would provide valuable information on the retention of transfer students in four-year programs and on the value of completing A.S. degrees prior to transfer. Policies and programs would be designed differently, depending on the answers to these questions.

The lack of information (especially longitudinal and comparative information that can be disaggregated by gender, race/ethnicity, and other background variables) on the successes and failures of students who begin their engineering educations in two-year and four-year programs presents serious problems for an analysis of the transfer function of community colleges. Most often, community colleges lose sight of students once they transfer to four-year institutions, precisely when they should begin tracking the educational and career trajectories of their students. Compiling and publicizing data on transfer students' success in obtaining B.S. or advanced engineering degrees would demonstrate the effectiveness of engineering studies in community colleges nationally and improve their recruitment rates or point to the need to strengthen community college programs.

Less specific factors (e.g., economic, geographic, political, and social factors) also influence the effectiveness of attempts to improve the recruitment, retention, transfer, persistence to degrees (A.S., B.S., and advanced degrees), and the diversity of students on the community college pathway to engineering careers. Compiling and analyzing data on these factors, which are sometimes crucial to decisions, require more financial and personnel resources than smaller schools can afford.

KEY CHALLENGE

Some useful information is being collected. For example, primary data are collected by the National Center for Education Statistics at the U.S. Department of Education, National Science Foundation, and American Association of Engineering Societies Engineering Workforce Commission, which collects data on student enrollments and degrees. All of these data are necessary for policy makers and institutions to facilitate and evaluate the transfer process. However, the comparability and periodicity of data have yet to be determined. In fact, we do not know how many community colleges offer engineering sciences programs! The key challenge is to collect information disaggregated by institution and student characteristics, that is, longitudinal information. The Department of Education's data are noteworthy examples of useful longitudinal data.

EXEMPLARY APPROACHES

Among the workshop participants who reported that their institutions do compile data on their students, four-year educational institutions were more likely than community colleges to undertake elaborate data collection and analyses and to gather quantitative as well as qualitative data.

University of California, Davis

A considerable amount of data on transfer students is collected at UCD, where a student information system integrates information associated with a student's admissions application, transfer coursework, and UCD academic record. The Student Affairs Research and Information Unit frequently surveys students regarding their experiences during orientation, advising, retention, and extracurricular activities. The research unit also conducts surveys at intervals of 1, 10, 20, and 30 years after graduation. Alumni survey data are then linked to student data.

The workshop representative from UCD reported that much of the data has not been analyzed, especially with respect to community college transfer students, although the university hopes to do so when resources become available. One study of the persistence of community college transfer students showed that 89 percent of engineering transfer students completed a bachelor's degree at UCD. Students who transfer and students entering as freshmen have virtually the same graduation rate.

Tidewater Community College/Old Dominion University/ Virginia Polytechnic Institute and State University

Community colleges usually do not collect data on their students, and when they do, it is more limited in scope than student data collected by four-year institutions and qualitative rather than quantitative in nature. The Tidewater Community College Office of Institutional Research collects information primarily from voluntary postgraduation surveys. The surveys included questions about the college to which the student transferred, satisfaction with the preparation for transfer, and satisfaction with the program and instructors at TCC. Information from these surveys is typically available upon request.

Virginia Polytechnic Institute and State University share data on transfer students at articulation conferences, including (1) the overall acceptance rate for transfer students and TCC students and (2) grades (anonymously) for TCC transfer students.

West Kentucky Community and Technical College

In 1999, West Kentucky Community (WKC) and Technical College implemented a student survey to evaluate its programs and prepare for ABET accreditation. The surveys include questions about student backgrounds (e.g., age, family educational attainment, access to computers at home); commuting requirements; degree objectives; scheduling needs/ preferences, reasons for choosing a collaborative program; and whether they plan to continue complete an A.S. degree at WKC and/or transfer to a four-year engineering program. The response rate for these annual surveys has been 20–40 percent, and the results have been used in the ABET self-study. Recent surveys have also been conducted by the mathematics department to determine the effectiveness of the calculus series. WKC also collects information on students' evaluations of their community college experience during advising sessions when students discuss course selections and career goals. Responses are not quantified, but certain incidents are documented for internal purposes.

CONCLUSION

Workshop participants whose institutions do not collect data on student outcomes cited cost as a barrier. Even schools that do collect data on students cited the cost of analyzing these data as a barrier to using them to evaluate the effectiveness of programs. Systematic data collection programs are needed to determine educational and career outcomes for students who begin and complete their educations in community colleges and for students transferring from community colleges to four-year programs.

A recent discussion, convened by the National Governors Association—an initiative that was created in hopes of influencing the impending reauthorization of the Higher Education Act and other major educational laws—explored ways to track students through elementary, secondary, and higher education, which would provide better data on dropout rates and other weaknesses in American education (Chronicle of Higher Education, 2005; Cunningham and Milan, 2005). Other information that panelists suggested would be useful were data on students receiving student aid, the number who need remedial work when they enter college, the number who enter college, especially part-time students who may not enroll directly after high school. The discussion explored ways to improve those stages of education by "aligning" the federal laws that govern them, an initiative the association announced in February, 2005.

The initiative has been opposed by groups citing privacy concerns. To address this issue, the state of Delaware uses randomly assigned numbers to track K–12 students' dropout rates. Delaware is working to extend the use of those numbers to track the progress of students who attend public and private colleges in the state.

Conclusion 6-1 A comprehensive, systematic strategy for data collection on educational and career outcomes for community college and transfer students would require leadership in the engineering profession and from funding agencies to define the most relevant data items, to encourage collaboration between two- and four-year educational institutions, to provide for privacy of students, and to develop vehicles for dissemination.

Conclusion 6-2 A meeting of data-collecting agencies (e.g., National Center for Education Statistics, National Science Foundation Division of Science Resources Statistics, American Association of Community Colleges, and American Society for Engineering Education) would be an ideal forum for developing criteria for collecting data from different kinds of institutions.

7

Report Summary

The community college pathway to engineering careers can be enhanced by improving the enabling mechanisms related to articula tion and transfer. Statewide and institution-to-institution articulation agreements must be more transparent to students and their parents, and they must be more encompassing to address the many exigencies community college students experience. Articulation agreements should reflect the differences in resources available to two- and four-year educational institutions and provide flexibility for students and transfer partners, such as resistance to shocks (e.g., changes in course numbering and content). These enhancements will only be made when communication and collaboration improves between two- and four-year educational institutions; four-year colleges and universities are in the best position to initiate more interactive relationships with their community college partners.

The retention rate and persistence to the B.S. degree of transfer students is likely to improve only when two- and four-year institutions have established partnerships that are driven by mutual interest and investment, as opposed to the personal interest and commitment of individual faculty members or administrators. For this to happen, two- and four-year educational institutions will need to see themselves as stakeholders in students' outcomes such as recruitment, retention, and persistence to the B.S. degree.

A more comprehensive discussion in the engineering community on the question of standardization versus flexibility in the A.S. and four-year lower-division curricula and on measuring student learning outcomes would be beneficial. If these issues can be resolved, students are likely to have a higher comfort level about their chances of transferring successfully and more confidence in quality assurances.

The general public, as well as many community college students and their parents, are not aware of the community college transfer mission. Students who do not realize they can obtain a four-year engineering degree through the community college pathway tend to set their sights too low. In addition, a lack of understanding of what engineers do, or worse, a negative image of engineers and the engineering profession, also contribute to students' decisions to end their secondary education with a certificate or A.S. degree or to elect another major if they plan to transfer to a four-year institution. Engineering professional societies and other key stakeholders must work together to generate positive, compelling messages about the opportunities for engineers to improve the quality of life and sustainability, domestically and globally.

To increase public awareness of the community college mission, state and national policy makers must address the issue of community college capacity. In California, for example, students were reported being turned away from community colleges due to lack of capacity. Increasing capacity will require increased funding for community colleges, as well as for community college and transfer students.

Accessible, reliable data about student and institutional outcomes would make it possible to prioritize and address many of the problems outlined in this report. Currently, however, not enough data are available on community college student educational pathways, and institutions represented at the workshop reported that they do not have the funds to collect and analyze data on students. As community colleges become more important in higher education in the United States, data will be necessary to evaluate both student and institutional outcomes and to answer the questions raised in this report and elsewhere about the relationship between articulation agreements and recruitment, retention, and persistence to the B.S. degree of community college transfer students.

The majority of workshop participants reported that community college transfer students perform as well academically and have comparable retention rates to those of students who began as freshmen in four-year engineering programs. However, there is little systematic data to support this claim. Many workshop participants from four-year programs noted that community college students' performance often falters temporarily following transfer, but improves as students regain their confidence and become acclimated to their new environments. Information on the effectiveness of programs/approaches to reducing the culture shock would be helpful.

Fortunately, many institutions are already working to create or strengthen some of the enabling mechanisms noted above. Greater publicity of their efforts and more research into approaches that smooth the transfer experience for community college students would be beneficial.

ISSUES FOR FURTHER RESEARCH

During the committee's information-gathering activities for this report, a number of unresolved issues came to light that could be addressed through further research

- identification of ways to improve the clarity, transparency, and accessibility to documentation in two-year/four-year institutional partnerships
- identification of ways of institutionalizing transfer partnerships and improving communication between transfer partners
- identification of competencies/learning outcomes (rather than course lists or credits)—for community college students and other students—required for upper-division engineering courses
- the success rate of transfers before and after completion of the A.S. degree
- documentation of performance outcomes related to recruitment, transfer, retention, and persistence to degrees in undergraduate engineering education
- determination of who should collect data relevant to transfers
- the impact of financial aid on enrollment, retention, and completion in community college engineering science programs
- assessment of different approaches to K–12 outreach programs
- the best ways to publicize the transfer mission of community colleges
- assessment of curricular content and pedagogy in mathematics
- synchronization of engineering science coursework and lower-division coursework in four-year B.S. programs to facilitate smoother transfers
- assessments of distance learning, especially for community college students, including a detailed survey of engineering courses available on line, data on quality (e.g., student outcomes), and comparative costs of distance education and traditional classroom teaching
- costs and benefits of recruiting industry experts to teach at community colleges
- the feasibility and desirability of standardized accreditation of community college programs

- data on the recruitment and retention of underrepresented minority students and women in science and engineering programs in two-year and four-year institutions
- success rate of women and underrepresented minority students who transfer from two-year to four-year programs

RESEARCH QUESTIONS

A surprising number of engineering graduates begin their studies in community colleges. However, large numbers do not necessarily translate into effectiveness. Based on expert testimony and workshop discussions, several broad, overarching questions on this topic arose:

- What is the attrition rate of students who begin their engineering studies at community colleges (including those who fail to transfer and those who fail to receive a baccalaureate degree)?
- How many community college graduates later attend four-year degree granting educational institutions? How many are admitted to Research I Institutions?
- How many obtain graduate degrees?
- What career paths do students with A.S. degrees follow?
- Do engineering students who begin at community colleges perform as well, better than, or not as well as other students? What factors influence their success rate?

Transfers

Different types of transfer partnerships have been developed for different reasons and from different starting points. Additional research would be helpful to identify the underlying characteristics of successful partnerships. The committee identified the following critical areas for further research:

1. Do engineering students who begin at community colleges perform as well, better than, or not as well as other students?
2. What factors influence their success rate?
3. Minimum GPA requirements for transfer students vary among four-year institutions. Are community college transfer students held to a higher standard than students who begin in four-year engineering programs?
4. What is an appropriate minimum GPA for transfer students?

Recruitment and Retention

1. To what extent are stakeholders (e.g., the public, industry) aware that an engineering degree can be pursued starting at the community college level? Are there regional variations in awareness and, if so, what strategies are effective in raising public awareness?
2. How can community colleges and four-year institutions attract more high school students to engineering?
3. What factors in the culture, student services, and learning environments of community colleges correlate with the successful completion of coursework and transfer to four-year engineering programs?
4. What factors in the culture, student services, and learning environment of four-year engineering programs correlate with the retention of transfer students through completion of the B.S. degree?
5. What can community colleges and four-year engineering programs learn from bridge programs? Can exemplary bridge programs be scaled up to improve recruitment and retention outcomes for both two- and four-year institutions?
6. What is the impact of financial aid on recruitment, retention, and transfer of community college students and on their retention to the B.S. and more advanced degrees?

Curricular Content, Quality, and Standards

1. What competencies should students have after two years in engineering science programs?
2. Should there be a common lower division curriculum, or should the curriculum be tailored to specific fields of engineering?
3. How can mathematics teaching be focused on engineering applications?
4. How can community colleges enlist industry engineers to share their skills and knowledge?
5. Can distance learning fill gaps in engineering science coursework at two-year programs?
6. What evidence is there that community college engineering science students learn effectively via online courses?
7. What impact would replacing in-person courses with online courses have on the engineering science infrastructure at community colleges?
8. Would online courses result in a decline in engineering science faculty? If so, how would this affect K–12 outreach activities, academic advisement, student clubs, and mentoring programs?

Diversity

1. Why is diversity among students and faculty important?
2. Is it necessary that students have same-race and same-gender role models?
3. What features of the academic environment are associated with recruitment and retention of women and minorities to engineering science and engineering programs?

Data Collection

1. What data would be useful for improving engineering pathways (e.g., data disaggregated by race/ethnicity or gender on transfer and completion rates)?
2. Who should collect important data?
3. How should data collection, especially by community colleges, be funded?

Although this study examines partnerships between community colleges and four-year engineering programs, the primary focus is on the needs of community colleges and their students related to articulation agreements and transfer processes. Further research is needed to better understand the perspectives of four-year educational institutions. Another direction for further research would be an in-depth examination of the experiences of a cohort of students entering and progressing through the community college pathway to an engineering career, using both quantitative and qualitative data collection methods.

Appendixes

A

Committee Members Biographical Information

JAMES M. ROSSER (*Chair*) is president of California State University, Los Angeles, and has been professor of health care management there since 1979. Before coming to Los Angeles, he was vice chancellor and acting chancellor of the New Jersey Department of Higher Education and associate vice chancellor for academic affairs and a tenured faculty member at the University of Kansas. Dr. Rosser serves on many civic and community boards, including the Los Angeles County Alliance for College Ready Public Schools, California Chamber of Commerce, Americans for the Arts, Community Television of Southern California (KCET), Los Angeles After-School Education and Child Care Program–LA's BEST, and Music Center Performing Arts Council/Education Council. His professional affiliations include the American Association of State Colleges and Universities, American Council on Education, Western Association of Schools and Colleges, Woodrow Wilson National Fellowship Foundation, California Council on Science and Technology, and numerous committees and commissions of the California State University system. He is a past chair of the Education and Human Resources Advisory Committee of the National Science Foundation and was chair of the National Academy of Engineering Forum on Diversity in the Engineering Workforce from 2000 to 2002. Dr. Rosser earned a B.A. and M.A. in microbiology and a Ph.D. in health education from Southern Illinois University at Carbondale.

ASHOK AGRAWAL is dean of Math, Science, Engineering and Technology and director of the Emerson Center for Engineering and Manufacturing at St. Louis Community College-Florissant Valley and a registered professional engineer in West Virginia. Prior to joining Florissant Valley, he was associate professor and chair of the Department of Engineering Technology at West Virginia University Institute of Technology. Mr. Agrawal is currently working toward the establishment of an advanced manufacturing center on the Florissant Valley campus. Mr. Agrawal's efforts have led to more than $1.5 million in improvements in the manufacturing laboratories at Florissant Valley in the past two years. In addition, he recently received a partnership grant, with the University of Missouri-Rolla, from the Society of Manufacturing Engineers Education Foundation to establish a 2 + 2 program in manufacturing engineering. He holds M.S. degrees in materials science and mining engineering from the University of Kentucky. From 1999 to 2001, he was a member of the National Research Council (NRC) Committee on Engineering Education, and from 1998 to 1999, he was a member of the NRC Board on Engineering Education.

WARREN J. BAKER has been president of California Polytechnic (Cal Poly) State University at San Luis Obispo since 1979 and is a registered engineer in four states. Prior to his appointment at Cal Poly, he was chief academic officer, vice president, Chrysler Professor, and dean of the College of Engineering at the University of Detroit. He was also a National Science Foundation visiting fellow at the Massachusetts Institute of Technology and a research associate at the U.S. Air Force Civil Engineering Research Facility at the University of New Mexico. In 1985, Dr. Baker was appointed by President Reagan to the National Science Board (NSB); he was reappointed in 1988 and served in a variety of leadership positions for 9 years. He is a founding member of the California Council on Science and Technology and the California Business-Higher Education Forum; a board member of the National Association of State Universities and Land Grant Colleges and chair of its Commission on Information Technologies; a member of the Business-Higher Education Forum and co-chair of its Math and Science initiative; a director of Westport Innovations Inc., and John Wiley and Sons Publishers; and a board member of the Society of Manufacturing Engineers Education Foundation. He was chair of the Board of Directors of the Civil Engineering Research Foundation, a nonprofit organization founded by the American Society of Civil Engineers to promote and fund research on the national infrastructure, U.S. competitiveness in design and construction, and the environment. Dr. Baker earned a B.S. and M.S. in civil engineering from the University of Notre Dame and a Ph.D. in geotechnical engineering from the University of New Mexico. He is a fellow of the American Society of Civil Engineers and the Detroit Engineering Society and the author of more than 35 papers on technology, distance learning, geotechnical engineering, risk analysis, and engineering education.

RICHARD CULVER, the Bartle Professor in Mechanical Engineering, Binghamton University, has been director of the Beta Coalition, a National Science Foundation-sponsored regional coalition of small engineering programs in central New York and Pennsylvania, and is active in research in learning theory and instructional models. He currently teaches introductory and senior-level design-project courses in engineering and works on curriculum development. As director of the Division of Engineering Design at Binghamton, he was responsible for coordinating the lower-division undergraduate program; he was also the first associate dean for academic affairs at the Watson School of Engineering and Applied Science at Binghamton University. For the past 20 years, he has been the liaison between Binghamton and community colleges in New York state; for the past 3 years, he has been an external reviewer of engineering science programs at eight campuses. Dr. Culver has taught at the Colorado School of Mines, University of Calgary, Ahmadu Bello University in Nigeria, and was a visiting professor at the University of Salford, England. He is a registered professional engineer in Colorado. He earned his Ph.D. from Cambridge University in 1964.

DAN G. DIMITRIU is the engineering coordinator for the Physics, Engineering, and Architecture Department at San Antonio College, the largest college in the Alamo Community College District of Texas. Prior to his faculty appointment at San Antonio College in August 2001, he was a practicing engineer for 32 years in the United States and in his home country of Romania. Dr. Dimitriu has been vice president of engineering for Prism Enterprises, which manufactures consumer products and automotive and medical equipment; a senior design engineer for Kinetic Concepts Inc., a medical equipment manufacturer; and director of engineering for Dynamic Industries Inc., a construction equipment manufacturer. Dr. Dimitriu is the author or coauthor of eight patents and has written technical papers on alternative fuels and engineering education, as well as conference papers on improving community college pathways in the development of a domestic engineering workforce. He received an M.S. in mechanical engineering from Polytechnic Institute of Bucharest, an M.B.A. from the Academy of Economic Sciences, Bucharest, and a Ph.D. in engineering from North Dakota State University.

JACK LOHMANN is associate provost at the Georgia Institute of Technology, where his principal responsibilities are the development, review, and accreditation of institutional academic programs and intercampus development of campuses in Metz, Paris, Singapore, and Savannah (Georgia). Dr. Lohmann has also held appointments at the National Science Foundation, University of Michigan, University of Southern California, and École Centrale Paris. His research and teaching interests are capital budgeting and economic decision analysis, and he has headed a number of engineering-education initiatives involving accreditation and curricular innovation. External sponsors of his research and educational initiatives include AT&T, GM, Hewlett-Packard, IBM, Microsoft Research, Motorola, National Science Foundation, Procter & Gamble, Sloan Foundation, and the United Engineering Foundation. He is currently editor of *Journal of Engineering Education*, a licensed professional engineer, and a fellow of the Institute of Industrial Engineers. Dr. Lohmann received a Ph.D. in industrial engineering from Stanford University.

MARGARET WEEKS is an adjunct program manager at ABET Inc. (formerly Accreditation Board for Engineering and Technology), where she is responsible for projects, workshops, conferences, and seminars in support of ABET's promotion of quality and innovation in engineering, engineering technology, applied science, and computing education. Ms. Weeks is also associate director of the Center for Technological Literacy at Hofstra University, where she is working with community colleges and technology educators to develop professional-development programs throughout New York state. Prior to joining ABET, she was a program director with the National Science Foundation (NSF) Division of Undergraduate Education and manager of projects in the NSF Advanced Technological Education and Course, Curriculum, and Laboratory Improvement programs. Between appointments at NSF, first from 1996 to 1998 and then from 1999 to 2001, Ms. Weeks taught courses in information and engineering technology at Charles County Community College (now the College of Southern Maryland) and was project director for a Microsoft/American Association of Community Colleges Working Connections Grant to improve information technology programs at the college. She earned an M.S. in ceramic engineering from Alfred University.

AARON WENGER, Professor Emeritus at Itasca Community College in Grand Rapids, Minnesota, has been a leader in establishing links between four-year degree-granting universities and two-year engineering programs at community colleges in Minnesota. In 1999, he was featured speaker at the University of Minnesota Articulation Conference, which brought together state universities and community colleges to discuss transfer issues for engineering. Dr. Wenger has focused on the role of community colleges in recruiting and retaining a cross-section of people in STEM-related careers, especially in the critical first years of a student's exposure to higher education and the demands of the engineering curriculum. In 2002, he was co-chair (with Ron Ulseth) of a national engineering-education conference at Itasca to discuss how community colleges could act as a bridge to higher education. Based on recommendations of conference participants, Professor Wenger is working on a $5-million grant request to the National Science Foundation to create Centers for Engineering Education Dissemination, which would bring together higher educational institutions, community colleges, degree-granting universities, and professional engineers in a national discussion on practices in engineering education. He has developed pedagogical practices that stress a strong learning community and early exposure to engineering design, which have resulted in the transfer of more than 80 percent of entering freshmen to degree-granting universities. Professor Wenger received a B.S. in engineering physics and an M.S. in astrophysics from the University of Toledo.

VERA ZDRAVKOVICH, vice president for instruction at Prince George's Community College in Maryland, supervises 254 full-time and more than 500 adjunct faculty members. During her tenure, she has established a Collegian Centers and an Honors Academy; implemented a comprehensive faculty professional-development program; developed and implemented a course and general education assessment model; reorganized the divisional and departmental structure of the college; reorganized instructional technology; established a new faculty Technology Resource Center and a new department of education; initiated the development of an integrated student-centered model; and directed National Science Foundation-supported institution-wide reform. In 2001, Dr. Zdrakovich was awarded the National Council of Instructional Administrators Instructional Leadership Award. She holds a Ph.D. in organic chemistry from George Washington University and a B.S. in chemical engineering from the University of Belgrade, Yugoslavia.

Project Liaisons

STEVE DIRECTOR is the Robert J. Vlasic Dean of Engineering and a professor of electrical engineering and computer science at the University of Michigan. He received a B.S. from the State University of New York at Stony Brook and an M.S. and Ph.D. in electrical engineering from the University of California, Berkeley. In 1977, he joined the faculty at Carnegie Mellon University, where he became the U.A. and Helen Whitaker University Professor of Electrical and Computer Engineering, head of the Department of Electrical and Computer Engineering, and dean of the College of Engineering (until June 1996). Dr. Director is a pioneer in computer-aided design and has a long record of commitment to and innovation in electrical and computer engineering curricula. He has published more than 150 papers and is the author or coauthor of six textbooks. Dr. Director is currently chair of the National Academy of Engineering (NAE) Committee on Engineering Education, chair of the Board of Directors of the American Society for Engineering Education Engineering Deans Council, and has served on numerous other boards and committees and as a consultant to industry and academia. He has received many awards for his research and educational contributions, including the 1998 IEEE Education Medal and the 1999 University of California, Berkeley, Distinguished Engineering Alumnus Award. Dr. Director is a fellow of IEEE and a member of NAE.

KARL S. PISTER is chair of the Governing Board of the California Council on Science and Technology, former vice president for educational outreach of the University of California (UC) System, and Chancellor Emeritus of UC, Santa Cruz. He began his 50-year career in higher education as assistant professor in the Department of Civil Engineering at UC Berkeley, where he progressed to chair of the Division of Structural Engineering and Structural Mechanics and then dean of the College of Engineering, a position he held for 10 years. From 1985 to 1990, he held the first Roy W. Carlson Chair in Engineering, and from 1991 to 1996, he was chancellor, UC Santa Cruz. Dr. Pister is a member of the National Academy of Engineering; a fellow of the American Academy of Arts and Sciences, American Academy of Mechanics, American Society of Mechanical Engineers, and American Association for the Advancement of Science; and an honorary fellow of the California Academy of Sciences. He is also a member of the Board of Directors of the Monterey Bay Aquarium Research Institute, the Center for the Future of Teaching and Learning, and the Board of Trustees of the American University of Armenia and the Monterey Institute of International Studies. Dr. Pister was founding chair of the National Research Council Board on Engineering Education.

RICHARD TAPIA, Noah Harding Professor of Computational and Applied Mathematics; associate director of graduate studies, Office of Research and Graduate Studies; and director of the Center for Excellence and Equity in Education at Rice University, is internationally known for his research in computational and mathematical sciences and for his leadership in education and outreach programs. He received his B.A., M.A., and Ph.D. in mathematics from the University of California, Los Angeles (UCLA) and is the author or coauthor of two books and more than 80 research papers. Dr. Tapia has delivered numerous addresses at national and international conferences and is a member of several national advisory boards. He has received many honors, including: the 2004 American Mathematical Society Distinguished Public Service Award and an honorary doctor of engineering degree from the Colorado School of Mines. He has served as chair of the National Research Council Board on Higher Education and Workforce, co-chair of educational outreach and training activities for the University of Illinois Supercomputer Center and the San Diego Supercomputer Center, and co-chair of the Research Board for Building Engineering and Science Talent.

B

Committee on Enhancing the Community College Pathway to Engineering Careers

April 1, 2004

Agenda

OPEN SESSION

8:30	Continental Breakfast (available in the committee room)
9:00	Welcome and Introductions, James M. Rosser
9:15	Comments on Charge from National Science Foundation *Panel:* Elizabeth Teles, Program Director, Education and Human Resources, NSF Bruce Kramer, Division Director, Engineering Directorate, NSF
9:45	National and State Perspectives on the Community College Transfer Mission *Panel:* Warren Baker, President, California Polytechnic State University, San Luis Obispo William Beston, Adjunct Professor of Physics, Texas A&M University-Corpus Christi
11:00	Break

11:15 Curricular Reform and Diversity Issues

Panel:
Ilene Busch-Vishniac, Professor, Department of Mechanical Engineering, Johns Hopkins University
Margaret (Peggie) Weeks, Program Officer, ABET

12:15 p.m. Lunch

1:00 Successful Community College-Engineering Educational Institution Partnerships

Panel:
Don Day, Professor of Engineering, Montgomery College, Rockville, Maryland
Aaron Wenger, Professor Emeritus, Itasca Community College

CLOSED SESSION

2:00 Committee Discussion
Deliberation of Panel Presentations
Discussion of Key Research Questions for July Workshop and Phase I Report

3:15 Committee's Next Steps
Review of Work Plan
Process for Selection of Participants for July Workshop

4:00 Adjourn

C

Workshop on Key Issues and Exemplary Practices in Community College Engineering Science Programs and Transfer

July 7–8, 2004

Agenda

Wednesday, July 7

8:30	Continental Breakfast
9:00	Welcome and Overview of the Workshop: The Importance of Enhancing the Community College Pathway to Engineering Careers James M. Rosser, President, California State University, Los Angeles, and Chair of the Committee on Enhancing the Community College Pathway to Engineering Careers John Slaughter, President and CEO, NACME William Wulf, President, National Academy of Engineering
9:30	Charge to the Study Committee and Purpose of Workshop: Perspectives of the National Science Foundation, Project Sponsor Bruce Kramer, Division Director, Division of Engineering Education and Centers
9:45	Keynote Address: Opportunities and Challenges in Improving the Community College Transfer Mission

Lucy Casale, Director of Community College Programs, Mathematics Engineering Science Achievement Program (MESA)
Sandy K. Brooks, Program Manager, Hewlett-Packard Philanthropy and Education

10:30 Break

10:45 Articulation, Transfer, and Curriculum: Perspectives of Two- and Four-Year Institutions
Moderator: Richard Culver, Committee Member
Presentations by:
- Itasca Community College and University of North Dakota
- Hudson Valley Community College and RPI
- Jones County Jr. College and Mississippi State University
- Montgomery College and University of Maryland
- Tidewater Community College and Old Dominion University

11:45 Break-out Sessions

12:30 p.m. Lunch (and Poster Sessions)

1:30 Key Issues in Enhancing the Community College Pathway
Moderator: James Rosser, Committee Chair
Susan Hackwood, Professor of Electrical Engineering, University of California-Riverside, and Executive Director, California Council on Science and Technology
Frank Hart, Dean, Educational Institution of Engineering and Computer Science, Bluefield State College, West Virginia
Jane Weyant, Assistant Dean, College of Engineering, Georgia Institute of Technology

2:30 Break-out Sessions

3:15 Break (and Poster Sessions)

3:45 Recruitment, Retention, Financial Aid and Diversity: Perspectives of Two- and Four-Year Institutions
Moderator: Aaron Wenger, Committee Member
Presentations by:
- Lenoir Community College and North Carolina State University

- Northern Essex Community College and Merrimack College
- Pima Community College and University of Arizona
- Prince Georges Community College and Howard University
- Three Rivers Community College and University of Massachusetts-Lowell

4:45 Break-out Sessions

5:30 Reception (and Poster Sessions)

Thursday, July 8

8:30 Continental Breakfast

9:00 Community College Transfer and Higher Education
Clifford Adelman, Senior Research Analyst, Department of Education

9:45 Standards and Quality: Perspectives of Two- and Four-Year Educational Institutions
Moderator: Peggie Weeks, Committee Member
Presentations by:
- Los Rios Community College and University of California-Davis
- Monroe Community College and Rochester Institute of Technology
- San Antonio College and University of Texas
- Highline Community College and Seattle University

10:45 Break

11:00 Break-out Sessions

11:30 Report Out to Plenary Session

12:15 p. m. Closing Remarks
James M. Rosser, Committee Chair

12:30 Adjourn

D

Workshop on Key Issues and Exemplary Practices in Community College Engineering Science Programs and Transfer

July 7–8, 2004

Attendees

Jack Adams‡
Chair, Department of Electrical and Computer Engineering
Merrimack College

Clifford Adelman†
Senior Research Analyst
Institute of Education Sciences
U.S. Department of Education

Ashok Agrawal*
Vice Chair, ASEE Two Year College Division
St. Louis Community College

Margaret Anderson‡
Assistant Dean
Rochester Institute of Technology

Sherry Anderson
Dean of Academic Affairs
West Kentucky Community and Technical College

J. Lyle Bagley‡
Dean of Engineering and Industrial Technology
Tidewater Community College

John Bailey‡
Associate Professor, Engineering Coordinator
Prince George's Community College

---

* = Steering Committee Member
+ = Project Liaison
† = Speaker
‡ = Presenter

Oktay Baysal‡
Dean, Frank Batten College of Engineering and Technology
Old Dominion University

Bill Beston†
Adjunct Professor of Physics
Texas A&M University-Corpus Christi

Mary Boleware‡
Physics Instructor
Jones County Junior College

Sandy K. Brooks†
HP Scholar Program Manager
Hewlett-Packard Philanthropy and Education

Lucy Casale†
Director of Community College Programs
Mathematics Engineering Science Achievement Program

Edward V. Clancy
Professor, Engineering Technology
California State Polytechnic University, Pomona

Bob Cornwell‡
Associate Professor of Mechanical Engineering
Seattle University

Richard Culver*
Bartle Professor in Mechanical Engineering
Binghamton University
State University of New York

Donald K. Day‡
Department of Physics and Engineering
Montgomery College

Martie De La Paz
Director of New and Transitioning Students
College of Engineering
Texas A&M University-Kingsville

Connie Della-Piana
Evaluation Consultant
Division of Undergraduate Education
National Science Foundation

Dan Dimitriu*‡
Engineering Program Coordinator
San Antonio College

Stephen W. Director (Chair, Committee on Engineering Education)+
Robert J. Vlasic Dean of Engineering
Robert H. Lurie Engineering Center
University of Michigan

Elizabeth M. Dorland
Program Director
Division of Undergraduate Education
National Science Foundation

---

* = Steering Committee Member
\+ = Project Liaison
† = Speaker
‡ = Presenter

David Finley
Dean, Allen Educational Institution of Engineering and Technology
Tri-State University

Gary Ford‡
Associate Dean for Undergraduate Studies
University of California, Davis

Bill Fortney
Director of Industrial Programs
Lenoir Community College

Terrence L. Freeman
Professor and Engineering Science Coordinator
St. Louis Community College

F. E. Gerlitz
Professor, Faculty
(INCOMPLETE?)
Washtenaw Community College

Paul Gordy
Associate Professor and Engineering Program Head
Tidewater Community College

Tim Grebner
Engineering Instructor
Century College

Robert A. Green‡
Undergraduate Coordinator
James Worth Bagley College of Engineering
Mississippi State University

Susan Hackwood†
Executive Director, California Council on Science and Technology, and Professor of Electrical Engineering
University of California, Riverside

E. Frank Hart†
Dean, Educational institution of Engineering Technology and Computer Science
Bluefield State College

Cheryl Hayden
Counselor
Glen Oaks Community College

Peter J. Hudleston
Associate Dean
IT Student Affairs
University of Minnesota Twin Cities

Jeffrey P. Jarosz
Baltimore, Md.

Muhammad Kehnemouyi‡
Professor and Chair
Montgomery College

Richard Keltie‡
Associate Dean for Academic Affairs
North Carolina State University
College of Engineering

* = Steering Committee Member
+ = Project Liaison
† = Speaker
‡ = Presenter

Bruce Kramer†
Director
Division of Engineering Education and Centers
National Science Foundation

Jack R. Lohmann*
Associate Provost for Institutional Development
Georgia Institute of Technology

Kenneth Manning‡
Department Chair, Mathematics and Engineering
Hudson Valley Community College

William Murphy
Director and Professor of Mechanical Engineering
University of Kentucky Extended Campus

Greg Ogden‡
Faculty, ATE Project Director
Pima Community College

Barbara Olds
Director, Division of Research, Evaluation and Communication
National Science Foundation

Zorica Pantic-Tanner‡
Dean, College of Engineering
The University of Texas at San Antonio

George Randy Parker‡
Vice President for Instruction and Student Services
Lenoir Community College

Laura B. Parr
Counselor and Associate Professor
Del Mar College

Mike Pelletier‡
Chair, Computer Technology and Engineering Department
Northern Essex Community College

Ken Rennels
Associate Dean for Undergraduate Education
Indiana University-Purdue University, Indianapolis

Ron Rockland
Associate Dean, Newark College of Engineering, and Associate Professor, Engineering Technology/Biomedical Engineering
New Jersey Institute of Technology

James M. Rosser (Chair)*
President
California State University, Los Angeles

Debbie Schatz
Assistant Director, Transfer Programs
University of Missouri at Rolla

---

* = Steering Committee Member
+ = Project Liaison
† = Speaker
‡ = Presenter

Roger K. Seals
Division of Undergraduate Education
National Science Foundation

Jim Sherrard‡
Chairman, Nuclear Program
Three Rivers Community College

John B. Slaughter†
President and CEO
National Action Council for Minorities in Engineering

Rebecca Sliger‡
Highline Community College

Richard Tapia (Chair, Board on Higher Education and Workforce)+
Noah Harding Professor
Rice University

Elizabeth J. Teles
Program Director
Advanced Technological Education
Division of Undergraduate Education
National Science Foundation

Jean-Louise Turner‡
Coordinator, Pre-College and Admissions
College of Engineering, Architecture and Computer Sciences
Howard University

* = Steering Committee Member
+ = Project Liaison
† = Speaker
‡ = Presenter

Ron Ulseth‡
Instructor of Engineering
Itasca Community College

John Wadach‡
Professor, Engineering Science and Physics
Monroe Community College

Jack Waintraub
Professor and Chairman
EET Department
Middlesex County College

John Watson‡
Dean, Educational institution of Engineering and Mines
University of North Dakota

Margaret Weeks*
Program Officer
ABET
Hofstra University

Aaron Wenger*
Professor Emeritus
Itasca Community College

Jane Weyant†
Assistant Dean for Undergraduate Studies
College of Engineering
Georgia Institute of Technology

Wm. A. Wulf†
President
National Academy of Engineering

Vera Zdravkovich*
Vice President for Instruction
Prince George's Community College

Staff

Mary Mattis
Study Director and Senior Program Officer
National Academy of Engineering

Lance A. Davis
Executive Officer
National Academy of Engineering

Peter Henderson
Director
Board on Higher Education and Workforce
National Research Council

Proctor P. Reid
Codirector, Program Office
National Academy of Engineering

John Sislin
Program Officer
Board on Higher Education and Workforce
National Research Council

Nathan Kahl
Senior Project Assistant
Program Office
National Academy of Engineering

---

* = Steering Committee Member
⁺ = Project Liaison
† = Speaker
‡ = Presenter

E

Transfer Data

The following data on transfer students has been collected (Kozeracki, 2001):

- gender, race, age, socioeconomic status, major or program, and full- or part-time status (often cross-tabulated with other findings
- student goals and aspirations
- impact of completing general education requirements or having to take developmental courses
- effects of special programs, such as honors programs, and choice of major on achievement
- grades (at the two-year and four-year institution)
- number of credits attempted and received
- withdrawal, persistence, graduation, and transfer rates
- time to degree

NATIONAL DATA

These data are available on three levels: national, state, and institutional.[1] In addition, some researchers have conducted one-time studies on particular questions, using primary research collection, such as student surveys. At the national level, sources of information include the National

[1]A review of data on transfer students (including an extensive reference section) can be found in Kozeracki (2001).

Science Foundation (NSF), U.S. Department of Education National Center for Education Statistics (NCES), and Transfer Assembly Project.

For the purposes of this study, NSF data on transfers are especially relevant because they focus on science and engineering. Information from graduate students regarding their experiences in community colleges is collected in the NSF National Survey of Recent College Graduates (NSRCG). "The survey questionnaire asks graduates who have received bachelor's or master's degrees in science or engineering fields whether they have ever attended a community college. The data collected do not distinguish between graduates who attended community college by taking one course and those who were enrolled full time (Tsapogas, 2004)." Because the survey also includes demographic questions, the findings can be disaggregated by marital status, race/ethnicity, age, and parents' education. In addition, some graduates were asked to give their reasons for attending community colleges. "Many S&E [science and engineering] graduates used community colleges for reasons other than to earn an associate's degree. In the 1999 survey cycle, 1997 and 1998 S&E graduates were asked why they decided to attend a community college. This question was not asked in the 2001 survey cycle" (Tsapogas, 2004).

Other sources of national-level data sources are useful for putting transfers to context but do not focus directly on engineering. NCES, for example, conducts a variety of surveys, including multiple surveys on postsecondary education, the most relevant of which is the Beginning Postsecondary Students (BPS) Longitudinal Study. Other surveys that include questions about attendance at community colleges are the Baccalaureate and Beyond Survey; National Postsecondary Student Aid Study (NPSAS), which focuses on financing a college education; National Longitudinal Study of the High School Class of 1972 (NLS-72); High School and Beyond (HS&B); and the National Education Longitudinal Study of 1988 (NELS:88). Analyses of these datasets for insights into the role of community colleges revealed the findings described below.

BPS includes several questions specifically about transfers. According to the NCES Web site (2004):

> [BPS] is designed specifically to collect data related to persistence in and completion of postsecondary education programs; relationships between work and education efforts; and the effect of postsecondary education on the lives of individuals. The current BPS Longitudinal Study is made up of people who first entered postsecondary education in the 1995–1996 academic year. These students were part of the National Postsecondary Student Aid Study (NPSAS). NPSAS includes everyone in postsecondary education, regardless of age or level of postsecondary enrollment. For BPS, students included in NPSAS who had just started their postsec-

ondary education were interviewed two additional times throughout their education and into the work force. The last interview took place in 2001 (NCES, 2004).

A final national data source on transfer students is the Transfer Assembly Project at the University of California, Los Angeles (UCLA) Community College Studies Program (2004).

> [The project] is the longest standing study focusing on statewide measures of transfers from community college to baccalaureate. Since 1989 the project has collected data on transfer rates using the following measure: The transfer rate is the percentage of all first-time community college students who complete at least 12 units at that college and who take at least one class at a public in-state university within four years of leaving the community college. Data are collected from individual institutions in a state, sometimes through the statewide agency, and are aggregated into a statewide rate, which is subsequently reaggregated into a national transfer rate. Because of confidentiality agreements, data are not published for individual institutions or for the states. Analysis of the changes in the rates, however, indicates that there are larger disparities in transfer rates between institutions in states than there are between states.

The Transfer Assembly Project assesses the transfer rate for students who enter two-year colleges with no prior college experience and complete at least 12 college credits in four years. This group is then compared to students who subsequently take one or more classes at a public in-state university in the next four years.

STATEWIDE DATA

At the state level, a number of entities, such as state higher education offices, collect data on student enrollments, including transfers, for institutions in the state (Welsh, 2002; Welsh and Kjorlien, 2001). Although many states collect some data, they have been slow to develop student-tracking databases. In addition, there are substantial differences among states in terms of the periodicity of collection and the scope of the surveys. Moreover, some of the data are not available on the web. A search of data that are available online suggested that much of the data cannot be disaggregated by gender, race and ethnicity, or field. Most important, research suggests that state higher education agencies do very little with the data they do collect (Welsh, 2002).

Some examples follow to illustrate the types of data collected by states. The California Postsecondary Education Commission collects data on full-year transfer information regarding the flow of community college students to four-year colleges and universities for the academic years between 1989–1990 and 2003–2004 (CPEC, 2004).

The state of Oregon has collected information since the 1995–1996 academic year on the number of students enrolled for credit at an Oregon community college one year and then at an Oregon university-system institution the next. Enrollment data are disaggregated by gender and race/ethnicity. Data are also collected on university majors declared by Oregon community college associate-degree-holders enrolled at Oregon university-system institutions the year after community college graduation; these data are aggregated by the community college degree and by university discipline. In addition, data are collected on the academic performance of undergraduate students—measured in grade point average (GPA)—enrolled at a community college in one year and at an Oregon university-system institution the next; the GPAs are compared to the GPAs of all students (Oregon University System, 2002).

INSTITUTION-SPECIFIC DATA

Individual institutions, both community colleges and four-year institutions, also collect transfer data. For example, Clark College, a two-year institution in Vancouver, Washington, released a transfer study in July 2002 based on a survey of 881 students who indicated on the registration form that they "intended to transfer" (Clark College, 2002). Students were grouped into three categories: (1) those who had just received associate's degrees; (2) those who had received associate's degrees and were still attending Clark; (3) and those who were no longer enrolled but were considered ready for transfer. The survey response rate was 66 percent (578 students). Of the 578 respondents, 411 (71 percent) transferred to four-year institutions (18 percent planned to transfer, 7 percent were not certain, and 4 percent would not transfer). About half of the students transferred to Washington State University. Engineering was the fourth-most common major among transfer students from Clark College.

SCOPE OF AVAILABLE DATA

Number of Students in Engineering Sciences and Engineering

We might address the question of how many students are in engineering sciences or engineering programs by looking at enrollment or the intention to major. Every year, the Higher Education Research Institute (HERI) conducts a survey of freshmen that includes a question regarding their intended majors. Of the freshmen who intend to major in S&E, the percentage that select engineering has been fairly stable, with a high of 36.3 percent in 1980 and a low of 25.8 percent in 1995; the average is about

31 percent. In 2002, the percentage of male freshmen intending to major in engineering was 44 percent of those intending to major in S&E. For women, the percentage was 11.5, down from the percentage of women in the 1980s and early 1990s. (More men selected engineering among the S&E fields; women tended to select biological or agricultural sciences or social or behavioral sciences over engineering, although engineering beat out the remaining sciences.) By ethnicity, the data suggest that the percentage of white freshmen who select engineering is relatively stable, between 8 and 10 percent. Asian Americans were more likely to select engineering during the 1980s. Interest among blacks and Mexican American and Puerto Rican American freshmen has increased slightly (NSB, 2004).

However, the HERI survey involves students at both community colleges and four-year institutions (primarily the latter). To separate these groups, the best strategy is to use NCES data. The NCES IPEDS database includes fall enrollments from all primary providers of postsecondary education, but these are not disaggregated by discipline.

A second way to identify the number of students in engineering is in terms of degree completion, for which NSF collects data. Table E-1 shows the number of associate degrees awarded in engineering sciences. Table E-2 shows the number of bachelor's degrees awarded in engineering.

As Table E-1 shows, the total number of associate degrees has not changed much for women since 1990; the number has declined for men. Because most associate degrees are awarded to white students, much of the decline can be explained by the decline in the number of white males who receive associate degrees. It should be stressed, however, that many students transfer prior to receiving associate degrees, so the potential number of transfer students in engineering is not evident from these data.

Table E-2 describes recent trends in the number of bachelor's degrees awarded. Since 1990, the number awarded to women has increased; the number awarded to men has remained relatively stable. At the bachelor's level, the number of white students receiving degrees has declined, while the number of minority students in all groups has increased. Enrollments are a less preferred measure, since they only include students who intended to major in engineering and succeeded in so doing at the end of their program. Not included are students who might have wanted to be engineers but dropped out of such programs at either the two-year or four-year institution.

Number of Transfers into Four-Year Engineering Programs

The group of students who could transfer from community colleges in engineering includes all students who are interested in pursuing this program, with or without a two-year institutional credential.[2] Once again,

TABLE E-1 Associate Degrees Awarded in Engineering, by Ethnicity and Gender, 1990–2001

Race & Ethnicity	Gender	1990	1991	1992	1993	1994	1995	1996	1997	1998	2000	2001
Total of All Races & Ethnicities	Female	288	279	355	330	362	307	269	271	323	238	297
Total of All Races & Ethnicities	Male	2,114	2,231	2,360	2,195	2,466	1,978	1,779	1,617	1,784	1,492	1,494
Black, Non-Hispanic	Female	15	20	31	28	31	28	24	24	24	22	15
Black, Non-Hispanic	Male	73	106	104	173	148	120	110	107	85	101	89
American Indian or Alaskan Native	Female	2	3	2	1	2	3	2	3	5	4	3
American Indian or Alaskan Native	Male	10	18	12	7	16	19	9	22	17	13	13
Asian or Pacific Islander	Female	27	18	21	12	36	24	14	20	25	23	25
Asian or Pacific Islander	Male	124	128	164	124	148	128	132	128	141	137	105
Hispanic	Female	21	26	24	43	32	25	19	10	28	28	32
Hispanic	Male	77	127	108	110	116	136	113	102	168	150	154
White, Non-Hispanic	Female	197	190	250	231	241	210	195	198	215	142	182
White, Non-Hispanic	Male	1,573	1,677	1,836	1,655	1,892	1,448	1,307	1,165	1,223	984	1,020

SOURCE: NSF, 2005.

TABLE E-2 Bachelor's Degrees Awarded in Engineering, by Sex, Race/Ethnicity, and Citizenship, 1990–2002

				U.S. Citizen/Permanent Resident					
Year	Total	Female	Male	White	Asian/ Pacific Islander	Black	Hispanic	American Indian/ Alaskan Native	Temporary Resident
1990	65,967	10,130	55,837	50,099	5,989	2,173	2,473	112	5,121
1991	63,986	10,016	53,970	48,028	6,305	2,304	2,663	146	4,540
1992	63,653	9,972	53,681	47,540	6,479	2,374	2,708	163	4,389
1993	65,001	10,453	54,548	47,976	6,764	2,637	2,845	175	4,604
1994	64,946	10,800	54,146	47,136	6,881	2,769	3,045	207	4,908
1995	64,749	11,303	53,446	46,264	7,056	2,897	3,409	230	4,893
1996	65,267	11,737	53,530	45,952	7,333	3,120	3,557	263	5,042
1997	65,091	12,160	52,931	44,976	7,625	3,203	4,005	265	5,017
1998	63,271	11,797	51,474	43,623	7,131	3,144	3,939	351	5,083
1999	62,500	12,360	50,140	42,650	7,226	3,171	4,073	328	5,052
2000	63,635	13,140	50,495	43,437	7,529	3,150	4,124	347	5,048
2001	65,195	13,195	52,000	44,407	8,340	3,182	4,152	275	4,839
2002	68,648	14,102	54,546	47,149	8,669	3,358	4,298	315	4,859

NOTES: Racial/ethnic breakouts are for U.S. citizens and permanent residents only. Temporary resident includes all racial/ethnic groups.
SOURCE: Engineering Workforce Commission, Engineering and Technology Degrees, 2002 (Washington, DC, 2002), cited in NSF (2004), Table C-8.

more data are available on general transfers than on transfers specifically in engineering. At the national level, one estimate places the transfer rate of students at around 25 percent (MacNeil, 2001).

Based on the NCES Beginning Postsecondary Students Longitudinal Study, we can make some general comments about transfers. For the 1989–1990 cohort of students assessed in 1994, the NCES study found, in general, that 25 percent of community college students indicated in their responses that they were working toward bachelor's degrees, and 39 percent of these students had transferred directly to four-year institutions by 1994. Of community college students identified as prospective transfers, those who were enrolled full time in their first year of community college were about twice as likely to transfer to four-year institutions within five years. Among community college beginners who transferred to four-year insti-

[2]A methodological note is important here. The transfer rate is the number of students who transfer from a two-year institution to a four-year institution divided by the number of students who could potentially transfer. There are different ways of defining both the numerator and denominator. See Bradburn and Hurst (2001).

tutions, 65 percent transferred without degrees; in other words, only about one out of three had completed associate degrees before transferring. One out of four community college transfers had received bachelor's degrees by 1994, while another 44 percent were still enrolled in four-year institutions, for an overall persistence rate of 70 percent. The bachelor's degree attainment rate was much higher among the minority of community college transfers who had completed associate degrees before transferring: 43 percent had received bachelor's degrees by 1994 compared with 17 percent of those who had transferred with no credential (McCormick and Carroll, 1997).

A second group of students who began at two-year institutions in 1995–1996, was also studied. This analysis showed that about one-quarter of students who started at community colleges said they intended to transfer to four-year institutions and complete bachelor's degrees. About one-half of those who intended to pursue bachelor's degrees did transfer (see Table E-3).

Adelman (1998) used the NCES Higher Educational Institution and Beyond Survey (which describes the activities of seniors and sophomores as they progress through higher educational institutions, postsecondary education, and into the workplace from 1980 through 1992) to examine the engineering pathway. The principal relevant findings are listed below:

- 20.1 percent of students who earned bachelor's degrees in engineering started in community colleges.
- One-third of students who reached the threshold of the baccalaureate engineering path, but who did not cross the threshold, started in community colleges.
- Of the students who left engineering, 14.8 percent started at community colleges (compared with 50.2 percent who started at comprehensive institutions and 25.8 percent who started at doctoral-degree-conferring institutions).
- The degree-completion rates (65.8 percent) of students who transferred from community colleges were equivalent to those of four-year-only college students (60.4 percent).
- Transfer students account for one-sixth of the degrees awarded in engineering.

The NCES Baccalaureate and Beyond Survey identifies bachelor's degree recipients for 1999–2000 by various characteristics, including first institution type and undergraduate major. Of students with bachelor's degrees in engineering, 17.8 percent began at two-year institutions. As Table

TABLE E-3 Percentage Distribution of Students Who Began at Public Two-Year Institutions in 1995–1996

Student Characteristics	% Distribution of Beginning Students	Percentage Transferred to a Four-Year Institution	Percentage of Transfers Who Completed an Associate's Degree First	Percentage of Transfers to Four-Year Institutions Who Persisted to a Bachelor's Degree		
					Persistence Status	
				Total	Completed Degree	Still Enrolled in Four-Year as of June 2001
Total (all beginning students)	100	28.9	33.3	78.9	34.7	44.3
Initial degree goal in 1995–1996: Bachelor's degree	24.8	50.8	18.8	82.3	44	38.3
Associate's degree	48.9	26.5	50.6	78.6	29.1	49.5

SOURCE: DOED, 2003.

TABLE E-4 First-Institution Type for Students Who Received Bachelor's Degrees in Engineering in 1999–2000

Undergraduate Major	First-Institution Type						
	Public Two-Year	Public Four-Year Nondoctorate Granting	Public Doctorate Granting	Private Not-for-Profit Four-Year Nondoctorate Granting	Private Not-for-Profit Doctorate Granting	Private For-Profit	Other
Engineering	17.8	8.0	50.4	6.5	16.3	0.7	0.3

SOURCE: Bradburn et al., 2003.

E-4 shows, a two-year institution is the second-most-frequent starting point for students who received bachelor's degrees in engineering.

Number of Students with Experience in Community Colleges

It is possible to survey students who received degrees at the bachelor's level or higher from the top down and ask whether they had experience in community colleges. According to data collected by the NSF, community college attendance among 1999 and 2000 (total) engineering bachelor's degree recipients was 44 percent; for master's degree recipients, it was 29 percent (NSF, 2003). The combined percentage of engineering bachelor's and master's degree recipients who attended community colleges was 40 percent (Tsapogas, 2004). The number of engineering bachelor's degree recipients who also earned associate degrees was 10 percent; the percentage for those who earned master's degrees was 8 percent (NSF, 2003). Together, the percentage of engineering bachelor's and master's degree recipients who received associate degrees in engineering was 22 percent (Tsapogas, 2004).

NSF also estimated data for students who received doctorates from 1991 to 1995; 8.2 percent of doctoral recipients in engineering had attended community colleges (NSF, 1996). Tsapogas (2004) updated this information for 1996–2000 (for all S&E) and again found that 8 percent had attended community colleges.

References

ABET (Accreditation Board for Engineering and Technology). 2004. CRITERIA FOR ACCREDITING ENGINEERING PROGRAMS. Available online at *http://www.abet.org/Linked%20Documents-UPDATE/Criteria%20and%20PP/05-06-EAC%20Criteria.pdf*. Accessed July 20, 2005.

AACC (American Association of Community Colleges). 2000. National Profile of Community College Trends and Statistics, 3rd ed. Washington, D.C.: Community College Press.

AACC and AASCU (American Association of State Colleges and Universities). 2004. Improving Access to the Baccalaureate. Washington, D.C.: Community College Press.

ACE (American Council on Education) and AAUP (American Association of University Professors). 2000. Does Diversity Make a Difference?: Three Research Studies on Diversity in College Classrooms. Washington, D.C.: ACE and AAUP.

Adelman, C. 1999. Women and Men of the Engineering Path: A Model for Analysis of Undergraduate Careers. Washington, D.C.: U.S. Government Printing Office.

Adelman, C. 2004. A Long Way Home: Tracking Potential STEM Students from High School through Community Colleges to . . . Presentation at the Enhancing Community College Workshop, July 7–8, 2004, Washington, D.C.: National Academies.

Bers, T. 1994. Articulation and Transfer. Pp. 247–261 in Managing Community Colleges: A Handbook for Effective Practice, edited by A. Cohen and F. Brawer. San Francisco: Jossey-Bass.

Beston, B. 2004. The Community College Transfer Mission. Presentation to the Committee on Enhancing the Community College Pathway to Engineering Careers, April 1, 2004, Washington, D.C.: National Academies

Bradburn, E.M., R. Berger, X. Li, K. Peter, and K. Rooney. 2003. A Descriptive Summary of 1999–2000 Bachelor's Degree Recipients One Year Later. NCES 2003-165. Washington, D.C.: U.S. Department of Education.

Bradburn, E.M., and D.G. Hurst. 2001. Community college transfer rates to four-year institutions using alternative definitions of transfer. Education Statistics Quarterly 3(3): 119–125.

CCST (California Council on Science and Technology). 2002. Critical Path Analysis of California's Science and Technology Education System. Riverside, Calif.: CCST.

Chronicle of Higher Education. 2005. Panelists Call for Tracking of Individual Students, Preschool Through College, for Better Education Data. Available online at *http://chronicle.com/daily/2005/06*.

Chubin, D.E., G.S. May, and E.L. Babco. 2005. Diversifying the engineering workforce. Journal of Engineering Education 94(1):73–86.

Clark College. 2002. Clark College Transfer Study 2000–2001. Available online at *http://www.clark.edu/pdf/general_information/transfer_study.pdf*. Accessed August 16, 2004.

CPEC (California Postsecondary Education Commission). 2004. CPEC Transfer Pathways. Available online at *http://www.cpec.ca.gov/OnLineData/TransferPathway.asp*. Accessed August 13, 2004.

Cunningham, A.F., and J. Milam (2005). *Feasibility of a Student Unit Record System within the Integrated Postsecondary Education Data System* (NCES 2005-180). U.S. Department of Education, National Center for Educational Statistics. Washington, D.C.: U.S. Government Printing Office.

Dimitriu, D., and J. O'Connor. 2004. Forging Stronger Ties between Community Colleges and Four-Year Universities. Proceedings of the 2004 American Society for Engineering Education Annual Conference and Exposition. Washington, D.C.: American Society for Engineering Education, 2-3. 6(1):41–56.

DOD (U.S. Department of Defense). 2001. Roadmap for National Security: Imperative for Change. Phase III Report of the U.S. Commission on National Security/21st Century. Washington, D.C.: U.S. Government Printing Office.

DOED (U.S. Department of Education). 2003. The Condition of Education 2003. NCES 2003–067. Washington, D.C.: U.S. Government Printing Office.

DOED. 2004. The Condition of Education 2004. NCES 2004–077. Washington, D.C.: U.S. Government Printing Office.

ECS (Education Commission of the States). 2001. Transfer and Articulation Policies. Denver, Colorado: ECS. Available online at *http://www.ecs.org/clearinghouse/23/75/2375.htm*. Accessed August 17, 2004.

EWC (Engineering Workforce Commission). 2004. Engineering and Technology Enrollments: Fall 2004, Washington, D.C.: EWC, American Association of Engineering Societies.

Grimson, J. 2002. Re-engineering the curriculum for the 21st century. European Journal of Engineering Education 27(1):31–37.

Heyman, GD., B. Martyna, and S. Bhatia. 2002. Gender and achievement-related beliefts among engineering students. Journal of Women and Minorities in Science and Engineering 8(1):41–52.

Hoachlander, G., A. C. Sikora, and L. Horn. 2003. Community college students: goals, academic preparation, and outcomes. Education Statistics Quarterly 5(2):121–128.

Ignash, J. 1999. In the shadow of baccalaureate institutions. ERIC Digest ED348129. November 17. Available online at: *http://www.ericdigests.org/1992-2/shadow.htm*.

Ignash, J. M., and B. K. Townsend. 2000. Evaluating state-level articulation agreements according to good practice. Community College Review 28(3):1–21.

Inman, W.E., and L. Mayes. 1999. The importance of being first: unique characteristics of first-generation community college students. Community College Review 26(4):3–22.

Jackson, S. 2002. The Quiet Crisis. Washington, D.C.: Building Engineering and Science talent (BEST).

Kasper, H. 2002–2003. The changing role of the community college. Occupational Outlook Quarterly 46(4):4–21.

Katsinas, S., J. Palmer, and T. Tollefson. 2004. State Funding for Community Colleges: Perceptions from the Field. National Council of State Directors of Community Colleges. Available online at *http://www.statedirectors.org/surveys/StateFIN.pdf*.

Kintzer, F. C., and J.L. Wattenbarger. 1985. The articulation/transfer phenomenon: patterns and directions. Washington, D.C.: American Association of Community and Junior Colleges.

Kozeracki, C. 2001. Studying transfer students: designs and methodological challenges. New Directions for Community Colleges (114):61–75.

Laanan, F.S. 2001. Transfer student adjustment. New Directions for Community Colleges (114):5–13.

MacNeil, W. 2001. Survey: long-stagnant student transfer rate jumps. Community College Week 13(18):2–3.

McCormick, A.C., and C.D. Carroll. 1997. Transfer Behavior among Beginning Postsecondary Students, 1989–94. ED 408 929. Washington, D.C.: U.S. Department of Education.

NAE (National Academy of Engineering). 2004. The Engineer of 2020: Visions of Engineering in the New Century. Washington, D.C.: The National Academies Press.

NCES (National Center for Education Statistics). 2004. Beginning Postsecondary Students (BPS) Longitudinal Study. Available online at *http://www.nces.ed.gov/surveys/bps/*. Accessed August 13, 2004.

NSB. 2004. *Science and Engineering Indicators 2004*. NSB 04-1 and NSB 04-1A. Arlington, Va.: National Science Foundation.

NSF (National Science Foundation). 1996. *Undergraduate Origins of Recent (1991-95) Science and Engineering Doctorate Recipients*, Detailed Statistical Tables. NSF 96-334. Arlington, Va.: NSF.

NSF. 2003. Characteristics of Recent Science and Engineering Graduates: 2001. NSF 04-302. Arlington, Va.: NSF.

NSF. 2004. Women, Minorities, and Persons with Disabilities in Science and Engineering: 2004. NSF 04-317. Arlington, Va.: NSF.

NSF. 2005. WebCASPAR: Integrated Science and Engineering Resources Data System. Available online at *http://caspar.nsf.gov/includes/checkJavascriptAbility2.jsp;jsessionid=F90C3CB9575FE5968E5551EB51AA9188?submitted=1*.

O'Connor, J., and D. Dimitriu. 2004. Getting an EDGE in Engineering Education. Proceedings of the 2004 American Society for Engineering Education Annual Conference and Exposition. Washington, D.C.: American Society for Engineering Education.

Oregon University System. 2002. Transfer Activity in Oregon Postsecondary Education, 1996–97 to 2000–01. Available online at *http://www.ous.edu/aca/OregonTransfer02.pdf*. Accessed August 16, 2004.

Pell Institute. 2004. Indicators of Opportunity in Higher Education. Washington, D.C.: Pell Institute.

Person, A., and J. Rosenbaum. 2004. For Needy Students, College Success Depends on More Than Access. Unpublished paper presented at the American Sociological Association meeting, August 15, 2004.

Phillippe, K.A., and L. Gonzalez Sullivan. 2005. National Profile of Community Colleges: Trends and Statistics. 4th ed. Washington, D.C.: Community College Press.

Rifkin, T. 1998. Improving Articulation Policy to Increase Transfer. Education Commission of the States Policy Paper, September. Denver, Colo.: Education Commission of the States.

Rovai, A. 2002. A preliminary look at the structural differences of higher education classroom communities in traditional and ALN courses. Journal of Asynchronous Learning Networks 6(1):41–56.

Seymour, E., and N. Hewitt. 1997. Talking About Leaving: Why Undergraduates Leave the Sciences. Boulder, CO: Westview Press.

SWE (Society of Women Engineers). 2004. NACME: Thirty years, thousands of stories. SWE Magazine 50(3):51–54.

SUNY (State University of New York) TYESA (Two-Year Engineering Science Association of New York State). 2005. Two Year Engineering Science Association of New York State. Available online at *http://www.tyesa.org*.

Tsapogas, J. 2004. The Role of Community Colleges in the Education of Recent Science and Engineering Graduates. NSF 04-315. Arlington, Va.: NSF.

UCLA Community College Studies. 2004. Recent and Ongoing Research Projects. Available online at *http://www.gseis.ucla.edu/ccs/research.htm*. Accessed August 11, 2004.

Ulseth, R., and A. Wenger. 2002. Follow-on Report of the Sugarlake Conference. Grand Rapids, Minn.: Itasca Community College.

Wellman, J. 2002. State Policy and Community College: Baccalaureate Transfer. San Jose, Calif.: National Center for Public Policy and Higher Education and the Institute for Higher Education Policy.

Welsh, J.F. 2002. Assessing the transfer function: benchmarking best practices from state higher education agencies. Assessment and Evaluation in Higher Education 27(3): 257–268.

Welsh, J.F., and C. Kjorlien. 2001. State support for interinstitutional transfer and articulation: the impact of databases and information systems. Community College Journal of Research and Practice 25(4):313–332.

Wulf, W.A. 2002. Diversity in Engineering: Managing the Workforce of the Future. Washington, D.C.: National Academy Press.

Wulf, W.A., and G.M.C. Fisher. 2002. A makeover for engineering education. Issues in Science and Technology 18(3):35–39.

Zamani, E.M. 2001. Institutional responses to barriers to the transfer process. New Directions for Community Colleges (114):15–24.